War, Politics, and Power

War, Politics, and Power

Selections from On War, and I Believe and Profess

Translated and Edited with an Introduction by

EDWARD M. COLLINS
Colonel, USAF

Karl von Clausewitz

With A New Foreword by Colonel Harry G. Summers, Jr.

Gateway Editions
Regnery Publishing, Inc.
WASHINGTON, D.C.

Library of Congress Cataloging-in-Publication Data

Clausewitz, Carl von, 1780–1831.
War, politics, and power : selections from On war and I believe and profess / Karl von Clausewitz ; translated and edited with an introduction by Edward M. Collins ; with a new foreword by Harry G. Summers, Jr.
ISBN 0-89526-401-3 (alk. paper)
p. cm.
1. War. 2. Military art and science. I. Collins, Edward M., 1917– . II. Title.
U102.C6623 1997
355.02—dc21 97-29173
CIP

Published in the United States by
Regnery Publishing, Inc.
An Eagle Publishing Company
One Massachusetts Avenue, NW
Washington, DC 20001

Distributed to the trade by
National Book Network
4720-A Boston Way
Lanham, MD 20706

PRINTED ON ACID-FREE PAPER.
MANUFACTURED IN THE UNITED STATES OF AMERICA

10 9 8 7 6 5 4 3 2 1

ACKNOWLEDGMENTS

As editor, I am naturally responsible for what this edition of Clausewitz' work includes, how it is stated, and what it omits. My work, however, rests on that of others, especially the professional soldiers and scholars whose study and research have kept Clausewitz's thought alive. Doctor Stefan T. Possony encouraged me to prepare the edition, assisted in checking difficult passages from *On War* in German, prepared all of the first translation of *I Believe and Profess,* and offered valuable suggestions and criticism of the first manuscript. His efforts made the edition better; the inadequacies which remain are, of course, my own.

EDWARD M. COLLINS

CONTENTS

FOREWORD

Although ignored by policymakers when it was first published in 1962, Air Force Colonel Edward M. Collins's *War, Politics, and Power*, a translated and distilled edition of Karl von Clausewitz's military classic, *Vom Kreige* (*On War*) (1832), came to have a profound impact on strategic thinking in the military and beyond. With Air Force Lieutenant Colonel David MacIsaac as its handmaiden, *War, Politics, and Power* became the catalyst for the introduction of Clausewitzian philosophy into the curriculums of all the service war colleges and led, in turn, to the inclusion of Clausewitz's theories in the military's post-Vietnam war-fighting doctrine. Indeed, it is not an exaggeration to say that *War, Politics, and Power* aided significantly in laying to rest the ghost of Vietnam and, eventually, to the overwhelming victories of the Gulf War. Clausewitz's *On War*, first in Collins's distillation and then in the full 717-page text, translated and edited by Michael Howard and Peter Paret (published in 1976 by Princeton University Press)[1], became the Rosetta Stone for the post-Vietnam military.

How could the writings of a Prussian veteran of the Napoleonic Wars, who knew nothing about America (an insignificant power at the time he wrote), have had such a powerful impact in this century on the United States, the world's sole superpower? The answer is that Clausewitz had a lot to tell us about fighting wars.

[1]Carl von Clausewitz, *On War*, translated and edited by Michael Howard and Peter Paret (Princeton, N.J.: Princeton University Press, 1976).

Thirty-five years ago, in his introduction to this volume—the first truly readable translation of Clausewitz—Colonel Collins asked the question, "Why read Clausewitz?" "The reason for reading Clausewitz today," he said, "is quite simple: he has something to say which is important, timely, and relevant to our situation." In 1962, the primary strategic "situation" was the Cold War between democracy and communism, and Collins assessed, correctly, the relevance of Clausewitz in terms of that external threat. He saw "close parallels between European conditions after the French Revolution in 1789 and world conditions since the Bolshevik revolution in 1917." Accordingly he devoted much of his introduction to Clausewitz's influence on communist conflict doctrine.

But, in 1962, when *War, Politics, and Power* was first published, the new and fashionable social-science theories of counterinsurgency were all the rage, and the century-old teachings of a Prussian militarist were dismissed as hopelessly quaint and out of date. Had Collins's *War, Politics, and Power* been read by Vietnam policymakers, it could have changed the course of history. As my Army War College text, *On Strategy: The Vietnam War in Context*[2], detailed almost two decades later, it was ignorance of Clausewitz's fundamental war-fighting principles that led to our failure in Vietnam. Like Cassandra, Collins's prophesies were correct. But, like Cassandra, no one would listen to him.

As it turned out, one person was listening—Collins's fellow Air Force officer David MacIsaac, who had received his doctorate in history while studying under the legendary Professor Ted Ropp at Duke University. Inspired by Collins's work and using it as a basic text, MacIsaac would lead a prolonged crusade to have Clausewitz's theories adopted as the U.S. military's basic teaching regarding the philosophy of war.

[2] Intended as a teaching text for the Army War College, *On Strategy: The Vietnam War in Context* was first published in April 1981 by the U.S. Government Printing Office. Under a new title and with a new epilogue, it was subsequently published commercially as well. (See note 14.)

It was an uphill battle from the start. Americans are not a particularly philosophical people to begin with, and the military is an even harder sell. By the very nature of their profession, the armed forces are especially action-oriented. Moreover, what war-fighting philosophy the military did have was a largely subconscious admixture of the mathematical formulas of Baron Antoine de Jomini's 1862 *The Art of War*[3] and the antipolitical views of Brevet Major General Emory Upton's posthumous 1904 work, *The Military Policy of the United States.*[4]

A contemporary of Clausewitz and a member of Napoleon's staff, Jomini commanded universal respect in the American military of the nineteenth century. In fact, Jomini's mathematical approach was so pervasive that it was often said that every officer on both sides in the Civil War went into battle with a sword in one hand and Jomini's *The Art of War* in the other. Reliance on his mathematical formulas become so popular that in 1869, in an address to the U.S. Military Academy, General William Tecumseh Sherman warned that "there exist many good men who honestly believe that one may, by the aid of modern science, sit in comfort and ease in his office chair, and with little blocks of wood to represent men, or even with figures and algebraic symbols, master the great game of war. I think this an insidious and most dangerous mistake."[5]

Ignoring this advice, General Emory Upton, a Civil War hero, taught that military and political affairs were entirely separate fields of knowledge. As a 1936 Army Command and General Staff College manual put it, "Strategy begins where politics ends." While there was always "the tendency for statesmen to meddle with the conduct of operations," the commander "must not allow himself to

[3] Baron Antoine de Jomini, *The Art of War* (Philadelphia: J. B. Lippincott & Co., 1862).

[4] Brevet Major General Emory Upton, *The Military Policy of the United States* (New York: Greenwood Press, 1968).

[5] William Tecumseh Sherman, *Address to the Graduating Class of the United States Military Academy* (New York: D. Van Nostrad Publishers, 1869), 8.

be paralyzed by such interference."[6] But as the Korean and Vietnam Wars were to demonstrate conclusively, Uptonian theory was not only out of touch with reality, but also flew in the face of the trinitarian nature of the U.S. military. The concept "trinitarian" comes from Clausewitz's observation that while in the eighteenth century war had been a matter for kings and princes, with the French Revolution war "suddenly became an affair of the people."[7] What emerged was a "strange trinity" of the people, the commander and his army, and the government. "A theory which left any one of them out of account...," Clausewitz warned, "would immediately contradict reality and be invalidated."[8]

Clausewitzian theory, which stood in direct opposition to the Jominian and Uptonian fallacies, was, however, virtually ignored. Following the lead of their British brethren, the American critics were resolutely anti-Clausewitzian, falsely identifying him with the German general staff and Teutonic aggression in the world wars. As Marine Corps Command and Staff College Professor Christopher Bassford has noted in his 1994 work, *Clausewitz in English: The Reception of Clausewitz in Britain and America 1815-1945*, "the dominating elements in mainstream British commentary on Clausewitz were ignorance and hostility.... The dominating fact about American commentary is that there was very little of it."[9]

After World War II not only was Clausewitzian theory on conventional war ignored, so too were all of the classic strategists. The common wisdom was that the advent of nuclear weapons had rendered all such theories obsolete. To some, as Temple University

[6] *The Principles of Strategy for an Independent Corps or Army in a Theater of Operations* (Fort Leavenworth, Kans.: The Command and General Staff School Press, 1936), 19-20.

[7] Karl von Clausewitz, *War, Politics, and Power*, translated and edited by Colonel Edward M. Collins, USAF (Washington, D.C.: Regnery Gateway, 1962), 227.

[8] Ibid, 86-87.

[9] Christopher Bassford, *Clausewitz in English; The Reception of Clausewitz in Britain and America 1815–1945* (New York: Oxford University Press, 1994), 191.

Professor Russell F. Weigley wrote, "The atomic bomb... represented a strategic revolution. The atomic explosions at Hiroshima and Nagasaki ended Clausewitz's 'the use of combats for the object of war' as a viable inclusive definition of strategy. A strategy of annihilation could now be so complete that a use of combats encompassing atomic weapons could no longer serve 'for the object of war.'"[10]

A prime example of this distorted view was the 1968 Penguin edition of *On War* edited by Anatole Rapaport, a professor of mathematical biology at the Mental Health Research Center at the University of Michigan. Warped by the anti-Vietnam War animus of the times and twisted by his political biases, Rapaport's analysis severely misinterpreted Clausewitz's philosophical insights on the relationship between war and society.

How insidious and how great a mistake lay in these anti-Clausewitzian theories would be proven by the ultimate Jominians, Defense Secretary Robert McNamara and his systems analyst "whiz kids," who, although never having heard of Jomini, nevertheless pursued Jomini's mathematical approach to the war in Vietnam—to the point of disaster. Even battlefield experience was declared of no value. As one civilian "whiz kid" told a senior combat veteran, "I've fought as many nuclear wars as you have." At the upper nuclear end of the conflict spectrum, military strategy was being made by what former Air Force Chief of Staff General Thomas D. White called the "tree-full-of-owls type of so-called professional 'defense intellectuals.'"[11] At the lower, limited war end, military strategy was being made by social and political scientists. They too preached that the Clausewitzian "use of combats" was obsolete.

[10] Russell F. Weigley, *The American Way of War: A History of United States Military Strategy and Policy* (New York: Macmillian Publishing Company, 1973), 365, 367-368.

[11] General Thomas D. White, USAF, quoted in Bernard Brodie, *War and Politics* (New York: Macmillian, 1973), 466.

"Limited war is an essentially diplomatic instrument," they preached. "Military forces are not for fighting but for signaling."[12]

In Vietnam the limited war theories deliberately excluded both the people and the military, and, as Clausewitz could have predicted, this mistake led to defeat on the battlefield.

These, then, were the theories that governed the conduct of the Vietnam War. But, as British strategist Correlli Barnett put it, the North Vietnamese were "terribly old fashioned" and were firmly committed to the Clausewitzian "use of combats for the object of the war." Their 1968 Tet offensive made it obvious that U.S. limited war theories were bankrupt and that America did not have a military strategy worthy of the name.

A theory of war was desperately needed, and Lieutenant Colonel MacIsaac, then on the faculty of the U.S. Air Force Academy, found the answer in Collins's *War, Politics, and Power.* In 1968, using Collins's work as the vehicle, MacIsaac succeeded in getting Clausewitzian theory incorporated into the 1968–1969 curriculum of the Air Force Academy. His crusade had taken off.

In 1975, granted a one-year military sabbatical from the Air Force Academy to the faculty of the Naval War College in Newport, Rhode Island, MacIsaac became part of the renaissance in military strategic thinking that would come to fruition almost two decades later in the Persian Gulf War. "Our increased reliance on civilians... to do our thinking for us reflected a failing of war colleges in general," Vice Admiral Stansfield Turner, a leader of this renaissance, said. "We must be able to produce military men who are a match for the best of the civilian strategists or we will abdicate control of our profession."[13] Spurning the false nuclear and limited war

[12] Stephen Peter Rosen, "Vietnam and the American Theory of Limited War," *International Security* (Fall 1982), 12-28.

[13] Vice Admiral Stansfield Turner, quoted in John B. Hattendorf, B. Mitchell Simpson III, and John R. Wadleigh, *Sailors and Scholars: The Centennial History of the U.S. Naval War College* (Newport, Rhode Island: Naval War College Press, 1984), 284-285.

prophets that had led the United States to disaster in Vietnam, Turner returned conventional war to center stage and with it long-neglected classic military strategists such as Alfred Thayer Mahan, Sir Julian Corbett, and Karl von Clausewitz. Working with the distinguished military historian, Professor Philip A. Crowl, chairman of the Naval War College's Department of Strategy, and again using *War, Politics, and Power* as the initial vehicle, Lieutenant Colonel MacIsaac was instrumental in integrating Clausewitzian philosophy into the Naval War College's 1975–1976 academic year curriculum.

The reemphasis on the fundamentals of conventional war spread from the Naval War College to the Air University at Maxwell Air Force Base in Alabama. Once again thanks to MacIsaac, who in 1979 had been reassigned from Newport to Maxwell, and with the aid of Air Force Lieutenant General Raymond B. Furlong, the Air University's commander, Clausewitzian theories were integrated into the curriculum of the Air War College. It is important to note that, by this time, Michael Howard and Peter Paret had published the complete text of *On War*, a sign of the academic recognition of the importance of Clausewitzian war-fighting doctrine.

Ironically, considering that *On War* is primarily concerned with land strategy, the Army was the last of the military services to join the Clausewitzian renaissance. It was not until 1981, under the leadership of Major General Jack N. Merritt, that *On War* was adopted as a teaching text at the Army War College at Carlisle Barracks, Pennsylvania. And once again MacIsaac played a crucial role. It was his discussions with faculty members at Carlisle, including my then officemate, Colonel Wallace Franz, that led to Clausewitzian philosophy being adopted by Army leaders.

In fact, it was those discussions that led to my own discovery of Clausewitz's insights into the nature of war. In 1979, disturbed by the failure of senior Army officers to come to grips with the Vietnam War, Army Vice Chief of Staff General Walter T. Kerwin, Jr., ordered me to Carlisle to examine what had gone wrong in Vietnam and to write a book explaining those failures—a book that could be used by the Army War College and possibly reach a civilian audience as well.

Faced with a mountain of material on the Vietnam War but lacking a unifying theme for analysis, I listened with growing interest to the conversations between Franz and MacIsaac on the timelessness of Clausewitz's expositions on military fundamentals. Turning in desperation to *On War*, I found, to my amazement, that the explanation for our Vietnam debacle had been laid out a century and a half earlier.

The United States had not lost the war because of the media or the politicians or because of wily Oriental stratagems. As I explained in the resulting book[14] (which a decade and a half later continues to be used as a teaching text by the military's war and staff colleges, and by many civilian universities), we lost the war because of a failure to appreciate and apply the basic principles of fighting wars laid out by Clausewitz 150 years earlier. The Vietnam War demonstrated a core truth of Clausewitz's theory of war: the necessity of recognizing the unbreakable bond between a people and its military forces. The greatest contribution Clausewitz would make to American strategic thinking would be in the realm of internal, not external, affairs. "Vietnam was a reaffirmation of the peculiar relationship between the American Army and the American people," said then Army Chief of Staff General Fred C. Weyand in 1976. The last U.S. commander in Vietnam, Weyand noted that "the American Army really is a people's Army in the sense that it belongs to the American people who take a jealous and proprietary interest in its involvement.... In the final analysis the American Army is not so much an arm of the Executive Branch as it is an arm of the American peo-

[14] *On Strategy: A Critical Analysis of the Vietnam War* is the commercial edition of *On Strategy: The Vietnam War in Context* (see note 1). Published in hardback by Presidio Press in Novato, California, in 1982, it was published in paperback by Dell Publishing in New York in 1984 and most recently republished as a trade paperback by Presidio Press in 1995. (See note 2.)

ple."[15] The U.S. military had inadvertently discovered its trinitarian roots.

Clausewitz had discovered in the French Revolution what the Founding Fathers had discovered earlier in the American Revolution and had institutionalized in the Constitution of the United States. "The power... of the British King extends to the Declaring of war and to the Raising and Regulating of fleets and armies," noted Alexander Hamilton in the Federalist Papers in March 1788, forty-four years before On War was published. "All which by the Constitution would appertain to the Legislature," he continued. Hamilton defined the legislature as "a popular body, consisting of the representatives of the people, periodically elected," thus explicitly making the U.S. military a "people's army" centuries before the term was coined.[16] Thus, strange as it may sound, Clausewitz's *On War* provides the perfect foundation for the American way of war. And that is what makes Clausewitz's work so valuable.

The complete text of *On War* is formidable, consisting of over seven hundred pages of small print divided into eight books, each consisting of several chapters. Much of the unabridged edition concerns what Collins calls "the craftsmanship of war, the tactical employment of forces of the kind then available." As he correctly observes, "it may be omitted, without loss, from an edition devoted to a broader view of war and strategy." Thus *War, Politics, and Power* is far more manageable for the lay reader; it draws primarily from Books One, Two, Three, and Eight of the original text. "The purpose of this little volume," Collins writes, "is to present the main, currently applicable thoughts of Clausewitz in brief but adequate form." In this endeavor he has succeeded admirably. "For the reader

[15] General Fred C. Weyand, "Vietnam Myths and American Military Realities," *CDRS CALL*, July-August 1976.

[16] Alexander Hamilton, "The Federalist, No. 69," in *The Federalist*, edited by James C. Cooke (Middletown, Conn.: Wesleyan University Press, 1961), 465.

who wishes to go beyond that aim, this edition will serve as a useful introduction to Clausewitz and as a summation of his concepts regarding the broader problems of war and conflict strategy."[17]

As a bonus, Colonel Collins added an afterword, "I Believe and Profess," drawn from Clausewitz's "Political Declaration" (Bekenntnissdenkschrift in German) written in February 1812, twenty years before *On War*.[18] Published in anticipation of Prussia's impending alliance with France, the complete work runs to some twenty thousand words. But this short excerpt captures the Clausewitzian passion that is absent in the more scholarly *On War*.

"Perhaps no better commentary on the importance of the moral element in war can be made than to recall that this non-rational argument... eventually enabled the Prussians to rise against France at the favorable moment and ultimately defeat Napoleon completely."[19] For the American reader, wedded by nature to hard facts and computerized analysis, it is well to remember that it was the failure of the intangible moral element, not the failure of our quantified physical military strength, that led to our defeat in Vietnam. Clausewitz could have told us so, but despite the best efforts of Colonel Collins, we were not listening. We ought not make that same mistake again.

Colonel Harry G. Summers, Jr., Infantry
United States Army (Retired)

[17] Clausewitz, *War, Politics, and Power*, trans. and ed. by Collins, 2-3.
[18] Carl von Clausewitz, "From the 'Political Declaration (1812),'" *Historical and Political Writings*, edited and translated by Peter Paret and Daniel Moran (Princeton, N.J.: Princeton University Press, 1992), 287-303.
[19] Clausewitz, *War, Politics, and Power*, trans. and ed. by Collins, 49.

INTRODUCTION

Why Read Clausewitz?

The reason for reading Clausewitz today is quite simple: he has something to say which is important, timely, and relevant to our situation.

At first it may appear surprising that concepts of war and strategy developed over one hundred thirty years ago should have real application to current problems. Reflection will show, however, that there are close parallels between European conditions after the French Revolution in 1789 and world conditions since the Bolshevik revolution in 1917. Clausewitz described the former period as one in which "war itself, as it were, had been lecturing." Would not the same description accurately portray our own times? Like the Europe of 1789–1815, the world since 1917 has been characterized by social upheaval, violence, revolution and war. Ours has been an age of continuous and open non-military conflict between powerful states during periods of nominal peace. It is marked by the rise of nationalism. It is an era of rapid change in the nature of warfare. Thus, the turbulent span of years during which Clausewitz lived, fought, studied war, and wrote was very much a preview of our own times. For that reason, much of the experience and thought which went into Clausewitz's writings has direct application to contemporary problems.

Moreover, Clausewitz devoted much attention to the higher direction of national strategy and the focusing and directing of all elements of national strength. This aspect of his work has received particular attention from Communist strategists, who have thoroughly studied Clausewitz and incorporated his views into their conflict doctrine. An understanding of Clausewitz's major ideas can thus lead not only to an improved conflict doctrine of our own, but to a better understanding of the doctrine of those who have ranged themselves against us.

On the other hand, Clausewitz's experience with ground force combat was so extensive that much of his writing was concerned with the craftsmanship of war, the tactical employment of forces of the kind then available. Although a considerable part of his comment in this field is still applicable, it has long been studied by the armies of all nations and has been thoroughly assimilated into their field service regulations. It may therefore be omitted, without loss, from an edition devoted to a broader view of war and strategy.

The purpose of this little volume, then, is to present the main, currently applicable thoughts of Clausewitz in brief but adequate form. For the reader who wishes to go beyond that aim, this edition will serve as a useful introduction to Clausewitz and as a summation of his concepts regarding the broader problems of war and conflict strategy.

Clausewitz's Life and Times

It is a truism that no man can be adequately understood outside the setting of the time and place in which he lived and his personal circumstances and fortune. This applies with particular force to Clausewitz, whose unique contributions to strategic thought are peculiarly the outgrowth of the Europe of his time, his own career, and the fortunes of his country during his lifetime. Born in Prussia in 1780, and dying in 1831, his life was lived in a Europe in which three powerful currents—philosophical, social-political, and military—were undermining existing political and social institutions. These currents mutually influenced each other, and in combination they dominated the practical and intellectual experiences of Clausewitz.

The intellectual spirit of late 18th century Europe was one which ran largely counter to existing institutions and long-standing beliefs. European writers, particularly in France, were engaged in powerful attacks upon Church and State, and upon the philosophical foundations which supported these institutions. The spirit of change, most pronounced in France, ranged throughout Europe, and was particularly strong in some of the south German states. The century between the birth of Lessing in 1729 and the death of Goethe in 1832—the year after Clausewitz's death—has been called "the most glorious period in German cultural history."[1] The principal theme of this era—expressed by Lessing, Herder, Schiller, Goethe, Kant and others—was the importance of the individual, of individual freedom, and of the full development of the individual personality. Although these humanist writers were non-political, and were motivated by profound conviction rather than by a conscious desire to reform the state or other institutions, their views illustrate the marked contrast between the intellectual climate and the existing political and social institutions.

Clausewitz was thus born into a Prussia which, although having only recently emerged as a first class power in Europe, was already intellectually outmoded. It had been raised to the first rank by the intrigues, wars and autocratic control of Frederick the Great, who had in turn capitalized upon the success of his predecessors in developing the Prussian army. The army had, in fact, become the pivotal element in the Prussian state, and all economic, social, and political institutions had been subordinated to maintaining the size and efficiency of the army. Prussia was a poor state and the size of its army could be maintained only by practice of the most rigid economies in all other state functions. To accomplish this, the state was rigidly organized and disciplined; its administrative machinery was developed to fulfill Frederick's

1. Koppel S. Pinson, *Modern Germany: Its History and Civilization* (New York: MacMillan, 1954), p. 12.

requirement that his government should be "as coherent as a system of philosophy, so that finance, policy and the army are coordinated to the same end: namely, the consolidation of the State and the increase of its power.[2]

Frederick, an absolute ruler, effectively discouraged independence of thought in his subordinates and the bureaucracy. The nobility was required to provide the officer corps for his army, and was in return granted extensive privileges. The relatively unprivileged bourgeoisie produced equipment and paid taxes to support the army, and the peasantry supplied its food and common soldiers. Thus, "the army molded the state to its need; it was now the principal obstacle to political or social change of any kind."[3]

It is against this background of an autocratic, rigidly stratified, and highly disciplined military state that one must view Clausewitz's early circumstances. Clausewitz, the son of a minor tax collector at Burg, near Magdeburg, was born in 1780. His father's post provided only a moderate income and, as Clausewitz was one of six children, the family was unable to give him a secondary education. The French Revolution, leading as it did to the War of the First Coalition in 1792, provided him the opportunity to enter the army in 1792.[4] He served in the Rhine campaigns of 1793 and 1794, receiving his commission at the siege of Mainz in 1793. Prussia withdrew from the war in 1795, and Clausewitz was then assigned to garrison duties. There he worked to improve his education, with such success that he was assigned in 1801 to attend the War Academy, a school in Berlin for young officers.

2. G. P. Gooch, *Frederick the Great: The Ruler, The Writer, The Man* (New York: A. A. Knopf, 1947), p. 296.

3. Gordon A. Craig, *The Politics of the Prussian Army, 1640–1945* (Oxford: Clarendon Press, 1955), p. 19.

4. Clausewitz was neither younger nor less well educated than most cadets drawn from the nobility, who entered the army at about the same age because much of the nobility was impoverished and because length of service was then considered more important to advancement than education.

The director of the academy was Scharnhorst,[5] a Hanoverian who had, also in 1801, been persuaded by the Duke of Brunswick to transfer his services to Prussia. Brilliant and largely self-educated, Scharnhorst was already famous as a writer on military subjects, having earned that reputation by consistent and persevering study of war. Scharnhorst evidently saw in Clausewitz one of his own kind, and offered him special help and encouragement. This marked the beginning of a lifelong association during which Scharnhorst profoundly influenced Clausewitz's development and his career.

In 1803, Clausewitz, upon Scharnhorst's recommendation, became aide-de-camp to Prince August of Prussia. In 1806, he accompanied Prince August to the war with France, participating in the ill-fated Jena campaign which led to the virtually complete collapse of the Prussian Army. He was wounded and, with Prince August, taken prisoner by the French and held in France and Switzerland until 1808.

Prussia's humiliation and downfall were as complete as the collapse of her army. Napoleon entered Berlin on October 27, after having thoroughly routed the Prussians at Jena and Auerstaedt on October 14 and subsequently taken, against negligible resistance, a number of garrisons which a resolute defender could have held for a long period. He proceeded at once with Prussia's humiliation: "with his own hands he desecrated the tomb of Frederick the Great at Potsdam, and sent off his sword and scarf to the Invalides; he scrawled obscene insults against the Queen Luise on the walls of her own palace; he demolished the obelisk on the battlefield of Rossback; he carried off to Paris the figure of Victory from the Brandenburg gate, and drove the Prussian Guards like cattle down the Unter den Linden. . . ."[6]

5. Gerhard Johann David von Scharnhorst (1755–1813).

6. J. A. R. Marriott and C. Grant Robertson, *The Evolution of Prussia* (Oxford: Clarendon Press, 1915), p. 217.

Worse was to come. The war was terminated in 1807 by the Treaties of Tilsit and Koenigsberg, which—with later additions of the Treaty of Paris in 1808—made Napoleon absolute master of Prussia. Briefly, these treaties reduced Prussia from 89,120 square miles to 46,032 square miles, required it to pay the enormous indemnity of 140 million francs, obliged it to support an occupation garrison of 150,000 French troops until the indemnity was paid, and imposed a number of other terms which left France clearly Prussia's master.

Even more significant than the magnitude of the disaster itself was the reaction of much of the Prussian nation to it. Although German intellectuals had at first welcomed the French Revolution, and even been enthusiastic regarding it, a reaction had set in because of the increase in violence as the revolution developed, the attacks on religion and on property, the influence of Burke's *Reflections on the French Revolution,* and the increasingly apparent channeling of revolutionary vigor into French imperialism. Nevertheless, in 1806, sympathy for the French and for Napoleon was still widespread in both intellectual circles and among the people. Hegel could write, on the day before the battle of Jena: "As I formerly, now everybody wishes success to the French army."[7] Moreover, after the Prussian defeat "the support which Napoleonic rule received from the German people was much more widespread than the impression subsequently created by Prussian writers."[8] Some of this support came from those who had consistently favored the French Revolution, but much of it derived from the apathy of the long disciplined Prussian people and the reforms initiated by Napoleon in the areas directly administered by him. In these areas, feudalism was abolished and land reforms in the former ecclesiastical states made way for small peasant proprietors. Freedom of worship was recognized, and Jewish disabilities were removed. The Code Napoleon was installed as the legal code. Freedom of trade was established, and

7. Quoted in Pinson, *op. cit.*, p. 33.
8. *Ibid.*, p. 32.

the French system of weights and measures introduced. These were positive reforms with clearly tangible benefits to many, and it is little wonder that they attracted support.

It is also little wonder that the leadership of a nation which had suffered so complete a disaster—and, on the whole, accepted it apathetically—should look to the validity of national institutions. This examination had in fact begun some years previously, when Hardenberg[9] and others had recognized the great increase in French unity and strength which had been brought about by the revolution, and had correctly judged that a state organized and directed in the manner of Prussia and the other states of Europe could hardly hope to survive in the new atmosphere. Hardenberg had written, in a memorial to the King in 1794:

> The French Revolution . . . has brought the French people a new vigor despite all their turmoil and bloodshed. . . . It is an illusion to think that we can resist the revolution effectively by clinging more closely to the old order, by proscribing the new principles without pity. . . . The force of these principles is such, their attraction and diffusion is so universal, that the state which refuses to acknowledge them will be condemned to submit or to perish. . . . Democratic rules of conduct in a monarchical administration, such is the formula. . . which will conform most perfectly to the spirit of the age.[10]

The needed reforms had, of course, not been enacted before the disaster of 1806 and there were many—like Count Lottum and Generals York and Knesebeck—who held even after Jena and Tilsit that basic reforms or reorganization were not required.

The story of the reforms which produced the regeneration of Prussia is one which cannot be detailed here, but must be men-

9. Karl August van Hardenberg (1750–1822).
10. Quoted in Pinson, *op. cit.*, p. 33.

tioned. It is significant because the reformers recognized that the military reforms needed to raise Prussian military capabilities to the French level would have to be preceded by the political, social and economic reforms which had produced the radical increase in French military power. The military reforms were carried out by a group headed by Scharnhorst, and including Gneisenau, Boyen, and Grolman. Upon his return to Prussia in 1808, Clausewitz was brought into this group by Scharnhorst and was thereafter active in the Prussian military reforms until 1812. This group recognized the dependence of military reform upon political reform, and consequently supported and even advocated the political reforms which were being sought by Hardenberg and others. The military was thus a leader in producing the reforms which were absolutely essential to Prussia's regeneration. However, Scharnhorst and his followers were opposed by a considerable number of influential military and political figures, who were later to gain influence with the King and greatly reduce the influence of the reformers.

The political and social reforms were begun by Hardenberg and, after Napoleon forced his dismissal in 1807, carried out by von Stein, to whom the King gave direction of all of Prussia's internal and external affairs. Although the reformers were resisted throughout, and never succeeded in carrying out all the measures they considered essential, their achievements undoubtedly were the major factor in the regeneration of Prussia after 1812. The non-military reforms included the abolition of the legal aspects of hereditary serfdom, the institution of local self-government in the cities, the granting to nobles of permission to engage in "citizen occupations," the elimination of old economic corporations and medieval economic regulations, and the establishment of complete freedom of occupation and of contract.[11]

These changes were far-reaching and permanent. Their political effect was the beginning of citizen participation in the government

11. *Ibid.*, p. 37.

of Prussia. Economically, they introduced complete freedom of occupation, and marked the beginning of free enterprise and the consequent release of initiative. Socially, they made possible a much greater degree of class mobility. In total, they made possible Prussia's emergence from a rigid autocracy to a more flexible, resourceful and modern state.

The military reforms included a new system of officer selection and advancement which abolished social preference and established a system of examinations as the basis for both selection and advancement. Although the conservatives and the King attached important reservations to this system, it nevertheless opened the officer corps to men with suitable educational qualifications, regardless of their class or social background. This measure was supplemented by a thorough reorganization of the military schools for officers and candidates for commissions. In addition, a higher level military school—which later became the Kriegsakademie—was founded in Berlin. A three year course for small groups of selected officers was given there, and the upper class of this group became the chief source of officers for the General Staff. New Articles of War were issued which softened the old Prussian system of brutal discipline and protected the individual soldier from arbitrary verdicts by his commander. French tactics and organization were carefully studied, new training manuals were issued, and a determined effort was made to improve the equipment of the army and the supply organization supporting it.

The most important change was probably the establishment in 1809 of a Ministry of War with general authority over "everything pertaining to the military. . . ."[12] Because of King Frederick William III's refusal (until 1814) to appoint a war minister, the ministry came under Scharnhorst's effective control until Napoleon forced his removal in 1811. In this position Scharnhorst was able to carry out measures which might otherwise have been difficult to effect.

12. Craig, *op. cit.*, p. 51.

The principle of universal military service, based on conscription, was also recognized by the King during the reform period, but he refused to put it into effect. The reformers' plans for a large mobilization potential, based on training of conscripts and placing them in a national militia, were thereby frustrated. The Kruemper system, under which each company or squadron gave annual leave to a number of trained men and replaced them with raw recruits, was devised as a partial substitute. Contrary to German patriotic beliefs, it was not highly successful: the Prussian army and its trained reserves numbered only 65,675 officers and men at the outbreak of war in 1813.[13]

The reform period was of major importance in Clausewitz's intellectual development. He was associated with the leading Prussian military and political figures of his day, and he was a participant in far-ranging discussions and studies involving the internal connection between domestic, foreign and military policy. His associates were, moreover, men of noteworthy intellectual attainments. Concurrent with the reform, there was an intellectual upsurge in which leading figures—such as the philosopher Fichte, the "political preacher" Schleiermacher, and the newspaper editor Goerres—sought to rouse Prussian patriotism. The extent to which any of these figures influenced Clausewitz is conjectural, although one is led to believe, by his later attempt to explore the abstract nature of war as a "thing-in-itself," that he was influenced by Kant's philosophy.

The influence of the reform group began to wane in 1809. It is reported that, during that year, Clausewitz considered entering the British or Russian service to avoid having to serve under Napoleon. However, his attachment to Scharnhorst and his judgment of where his duty lay were undoubtedly factors holding him in Prussia.

The events of 1809, 1810, and 1811 moved the reform group to despair. The King appeared to have lost hope for Prussia's eventual resurgence, for he denied the appeal of the patriots to rise to oppor-

13. *Ibid.*, p. 50.

tunities for action against Napoleon's rule. The climax was reached in March, 1812, with the ratification of the Franco-Prussian Treaty, which threatened the undoing of all the steps taken since 1807 to strengthen the army. The terms of the treaty provided that, in the event of a French war with Russia, Prussia was to provide the French with an auxiliary army of 20,000 men. No mobilization or troop movements in Prussia were to be conducted without Napoleon's consent; French occupation was to be restored, and two of Prussia's fortresses were to be immediately occupied by French troops.

The reaction in the Prussian army was one of complete disgust. Three hundred officers, constituting almost a fourth of the officer corps, submitted their resignations.[14] Boyen and Clausewitz were among this group, and Clausewitz wrote "I Believe and Profess," the ringing denunciation of the treaty and appeal to Prussian patriotism which appears in this edition as the final selection from his work.

With Boyen and a number of other leading Prussian officers, Clausewitz then took service with the Russians against France and Prussia. He was a major figure in the long retreat to Moscow, and later was the principal figure in negotiating the defection of the Prussian corps from the French. This event made the reversal of the French-Prussian alliance possible and marked the climax of Napoleon's disastrous invasion of Russia. Perhaps more than any other single factor, it led to Napoleon's ultimate defeat.

Clausewitz's participation in this campaign clearly exerted a major influence on his later views. Moreover, this episode in his life stands as one of the genuinely remarkable experiences of war and international conflict. The experience therefore deserves more than passing attention. Clausewitz writes as follows of the opening events in the campaign:

> In February of 1812, the alliance between France and Prussia against Russia was concluded. The party in Prussia, which

14. *Ibid.*, p. 58.

still felt courage to resist, and refused to acknowledge the necessity of a junction with France, might properly be called the Scharnhorst party; for in the capital, besides himself and his near friends, there was hardly a man who did not set down this temper of mind for a semimadness. In the rest of the monarchy nothing but a few scattered indications of such a spirit were to be found.

So soon as this alliance was an ascertained fact, Scharnhorst quitted the center of government, and betook himself to Silesia, where, as inspector of fortresses, he reserved to himself a sort of official activity. He wished to withdraw himself from the observation of the French, and from active co-operation with them, utterly uncongenial to his nature, without entirely giving up his relations to the Prussian service. This half measure was one of eminent prudence. He was able, in his present position, to prevent much mischief, particularly as regarded concessions to France in the matter of the Prussian fortresses, and he kept his foot in the stirrup, ready to swing himself into the saddle at the favorable moment. He was a foreigner, without possessions or footing in Prussia, had always remained a little estranged from the King, and more so from the leading personages of the capital; and the merit of his operations was generally at this time much exposed to question. If he had now entirely abandoned the service, it may be questioned whether he would have been recalled to it in 1813.[15]

Clausewitz then goes on to say that von Boyen, "an intimate friend" of Scharnhorst, Colonel von Gneisenau, and "several others among the warmest adherents of Scharnhorst, and of his political views. . .among whom was the author" left the Prussian service

15. Clausewitz, *The Campaign of 1812 in Russia* (London: John Murray, 1843), p. 1, 2.

to go to Russia. Clausewitz asserts that this was done with the King's consent.[16]

It will be apparent that Scharnhorst and the other reformers had adopted a position which, although technically legal, was in fact aimed at defeating the policy of King Frederick William III and his ministers. The position of the entire group was one of great delicacy. Their aim was the defeat of France, Prussia's ally, but they necessarily had to aim simultaneously at avoiding an accompanying defeat of Prussia. These purposes had to be accomplished under extremely difficult political and military circumstances. In less complex circumstances, those who took service with the Russians would have been regarded generally as traitors, and it is clear that they were so regarded by a considerable segment of Prussian officialdom, during and long after the 1812–1815 period.

Clausewitz went first to Vilna, the headquarters of Emperor Alexander I of Russia and also of General Barclay deTolly, commander of the First Army of the West and also War Minister.[17] A number of Prussian officers, including Gneisenau and Count Chasot, were already there. At this early stage, while the French attack was being awaited, there was much justifiable apprehension in the Russian headquarters. The situation confronting Russia was anything but promising. The effective strength of the Russian armies which could be brought into opposition to the French "was 180,000 strong, if taken at a high estimate; the enemy, at the lowest, 350,000, and Bonaparte their leader."[18] Gneisenau believed that the only hope lay in the enormous difficulty of the enterprise for the French, but left for England in the belief that he could do little in Russia and that everything possible should be done by England, Sweden and Germany to effect a diversion in the French rear.[19] The

16. *Ibid.*, pp. 2, 3.
17. *Ibid.*, p. 3.
18. *Ibid.*, p. 11.
19. *Ibid.*, p. 4.

idea held by Scharnhorst and others who had remained in Prussia was that the French must fail if Russia's great area were used effectively, its resources husbanded until the last minute, and no peace accepted by the Russians under any conditions. This idea was transmitted to the Russians by Count Lieven, who had been Russian minister in Berlin and who reached the Russian headquarters in Vilna some time after Clausewitz.

In spite of the general similarity of this concept to the course which the campaign actually took, Clausewitz's account makes it clear that actual events were the accidental result of adoption of a completely different concept. Phull, a Prussian colonel who had been several years in the Russian army and had reached the rank of lieutenant general but had not learned Russian, was used by the Russians to develop broad concepts for the defense of Russia. His plan, which called for withdrawal a considerable distance to a prepared position for a defensive battle at Drissa on the middle Dvina, had been tentatively accepted by the Russian command. Clausewitz and others inspected the Drissa position, judged it as one which invited disaster, and influenced the Russian leadership to continue the withdrawal beyond Drissa. Clausewitz then performed many of the surveys in search of a favorable defensive position, and his account establishes that inability to locate such a position, rather than any grand design, was primarily responsible for the long withdrawal. The relationship between this experience and Clausewitz's later firm conviction that the defensive is inherently the stronger form of warfare is quite clear.

After the Russian evacuation of Moscow, Clausewitz was transferred, by an order of the Emperor, to be chief of the general staff to the garrison at Riga. Upon his arrival at Petersburg, en route to Riga, Clausewitz learned of the French retreat from Moscow. At his own request, he was then assigned to the army of General Wittgenstein, which opposed the French and Prussian armies on the northern flank of Napoleon's retreat. He joined Wittgenstein some time after the middle of October, and served with him until the end of the campaign. Acting as emissary for Wittgenstein's "operations officer," General Diebitsch—a Prussian who had served in Russia throughout his career—Clausewitz engaged in a series of covert negotiations with General York, the commander of the Prussian corps opposing Wittgenstein, which led to the defection of the Prussian corps from the French cause.

The Clausewitz-York negotiations produced the famous "Convention of Taurroggen" of December 30, 1812. Under terms of this convention, the Prussian corps was declared neutral, and a district assigned to it as neutral ground in what is now the Lithuanian SSR. If either sovereign rejected the agreement, the Prussians were assured of a free march home in the shortest direction. If the King of Prussia rejected the convention, the Prussians engaged not to serve against Russia for a period of two months. Upon conclusion of this agreement, York wrote to the King of Prussia: ". . . Now or never is the moment when Your Majesty may tear yourself from the extravagant demands of an ally, whose intentions toward Prussia, in the event of his success, were involved in a mystery which justified anxiety."[20a]

Before Frederick William III could act to nullify this action, Stein arrived in East Prussia from Russia, "arranged with York for the summoning of an East Prussian LANDTAG, and persuaded that body to mobilize a Landwehr of all able-bodied men between 18 and 45. . . ."[20b] As the magnitude of the French disaster unfolded, Prussia's leadership recognized the opportunity it afforded. The King then transferred his headquarters from Berlin to Breslau and appointed a committee, under Scharnhorst's domination, to increase the army's strength as rapidly as possible. The standing army was immediately brought to full strength, universal conscription was initiated, and the formation of volunteer detachments from the propertied classes begun. On 16 March 1813 Frederick William

20a. *Ibid.*, p. 218ff.
20b. Craig, *op. cit.*, p. 59.

III declared war on Napoleon, issued the famous address "To My People" explaining the sacrifices which were to be demanded of them, and announced the creation of a Landwehr on the East Prussian model, to comprise all men 1840 not serving in the regular army or volunteer detachments. A month later, a similar edict created the Landsturm, an organization responsible for home defense and guerilla operations in case of need. The reformer's concept of the "nation in arms" thus became a reality.[21]

The Prussian force was quickly doubled as a result of these measures, and was tripled within a year. In 1813 Prussia sent 280,000 men into the field; a remarkable number in spite of the fact that first line, regular forces numbered only 68,000 of this total.

The precise role which Clausewitz played during the period 1813–1815 is not entirely clear. As a Russian officer, he superintended the formation of the Landwehr in East Prussia in 1813. He is reported later to have served as chief of staff to Count Wallmoden, and he served as a Russian liaison officer on the staffs of Scharnhorst and Bluecher for at least part of the 1813–1814 period. He reentered the Prussian service in 1814 and became chief of staff of a Prussian army corps, under Thielmann, in 1814 and 1815.

Scharnhorst died of wounds in 1813, and was replaced as Chief of the General Staff by Gneisenau. Capitalizing on the aroused patriotic fervor of the Prussian forces, and drawing upon Napoleon's strategy, Gneisenau brought the work of the reformers to fruition by developing in the Prussian army an offensive spirit which aimed constantly at the destruction of the enemy's forces. As later codified by Clausewitz, this concept—that the proper military objective is destruction of the enemy's forces—was to dominate Prussian military thought throughout the nineteenth and into the twentieth century, and to influence greatly all of the world's armies.

In the period of reaction which followed the Congress of Vienna, the Prussian reformers were regarded with suspicion and

21. This paragraph is drawn from Craig, *op. cit.*, p. 59ff.

mistrust by much of the Prussian political leadership and the hereditary nobility. Moreover, their reforming zeal sometimes overwhelmed their better judgment, and their enemies were thus able, by 1819, to bring about the removal of all from positions of major influence. Clausewitz, in 1818, was promoted to major general and appointed director of the War Academy in Berlin. This post was purely administrative, with no influence upon the curriculum or the methods of teaching. Clausewitz thereupon began *On War,* which he had begun to formulate in 1816–1817. He devoted the next twelve years, during which he remained as director of the War Academy, to writing it. In 1830, upon the outbreak of the Polish Insurrection, he was appointed chief of staff to Gneisenau, who commanded the Army of Observation at Posen. War did not come, and Clausewitz returned to Berlin in 1831. Shortly after his return—and like Gneisenau, Hegel, and many other prominent Prussians—he died of cholera.

Before his departure for Poland, Clausewitz had sealed the separate books which constitute *On War* in separate packets, evidently in preparation for his possible failure to return. This climactic work of a lifetime of study and practice of war was edited and published by his wife, assisted by his friends, shortly after his death. The accompanying notes left by Clausewitz make it clear that he did not regard the work as finished, but he nevertheless felt that "the leading features which predominate in these materials I hold to be the right ones in the view of war." He had intended one more revision, in which he would have made a more careful distinction between "the two kinds of war," which he described as the one which aims at "the overthrow of the enemy" and the other as one which aims only at making some conquests on the frontier of his country. He held that these two kinds would continue to blend into each other, but that their characters were wholly different and should accordingly be kept in separate categories. A further aim in the intended revision was to "expressly and exactly establish the point of view . . . from which war is regarded as nothing but the *continuation of state policy by other means.*"

Clausewitz's principal experiences were those of the period between his entry into the Prussian service in 1792 and his assignment as director of the War Academy. This was an almost continuous period of war, intensive diplomatic maneuvering between states, social and political change and reform, and intellectual ferment concerning the nature and end of man himself and the kind of political and social institutions which best accord with man's nature and purpose. Because the successful conduct of war depended upon success in acquiring powerful allies and in preventing one's prospective opponent from effecting a more powerful alliance, diplomacy and political warfare became almost integral with war itself. Similarly, the military changes which the French Revolution had made possible in France—particularly the *levée en masse* and the French spirit which permitted marches and maneuvers which had been inconceivable for the traditional armies of the times—had shown the connection between social and political reform and the possibility of military reforms and changes in doctrine and tactics. It is not surprising, therefore, that Clausewitz should recognize the intimate connection between foreign policy, domestic policy, and military policy which he expressed in his concept of war as the continuation of politics by other means.

He had, moreover, seen in the methods of Napoleon the tendency toward "absolute" war, in which a "nation in arms" makes an all out effort to overthrow an enemy nation, with the means to this end being the destruction of the enemy's military force in a decisive battle. At the same time, he had observed that reasons of policy frequently dictated that a war not become "absolute," that it be kept under control both as to its objective and the degree of violence employed to attain it. His concept of "two kinds" of war is therefore directly traceable to his experience and observation of warfare.

As previously noted, Clausewitz regarded the defensive, properly understood, as the stronger form of warfare. This view applies, he held, to both political and military conflict. This point of view was particularly troublesome to Clausewitz's later German disciples, who wished to stress the value of the offensive spirit and the decisive battle to destroy the enemy's forces, and it is by no means universally accepted today. Whatever one may think of its current or universal validity, it is worth remembering that Clausewitz's personal experiences supported his conclusion. He had observed how difficult it was to alter the political status quo, in his work during the Prussian reforms and in seeing the establishment, after the Congress of Vienna, of the "Metternich system" to forestall internal political changes in European monarchies. His view on the strength of the military defensive appears to be directly attributable to his experiences in Russia in 1812.

We cannot say—nor, probably, could Clausewitz himself—how much any or all of his experiences and the climate of opinion of his time shaped his conclusions. The examples above are cited merely to demonstrate the point that Clausewitz's strategic thought is more fully understood when the circumstances and events out of which it appears to have emerged are known.

Clausewitz's Influence on Communist Conflict Doctrine

Clausewitz has been very carefully studied in the Communist world, and his concept of the nature of war and politics and many of his strategic ideas have been incorporated into Communist doctrine and strategy. Lenin studied Clausewitz's works in late 1914 and early 1915.[22] This study was undertaken, after Lenin had studied Hegel's *Logic,* as an application of Hegel's dialectic method. Clausewitz was regarded by Lenin as an Hegelian, and Lenin saw Clausewitz's work as a practical application of Hegel's dialectical method. Lenin was greatly influenced by Clausewitz, and his own

22. The account which follows is drawn largely from Stefan T. Possony, *A Century of Conflict* (Chicago: Henry Regnery, 1953), p. 20ff.

tremendous influence on Communist thought has indelibly imprinted Clausewitzian concepts in the communist operational planner's mind. When Lenin went into hiding after the July uprisings of 1917, he took with him only two books—Marx's *The Civil War in France* and Clausewitz's *On War.*[23]

Although Lenin left no systematic notes from Clausewitz, his voluminous annotated extracts from *On War* clearly indicate the nature of Clausewitz's influence upon his own thinking. The concept most emphasized was the connection between war and politics, and particularly the concept that war is the continuation of politics by other means. "Lenin's notes emphasize that war is not only a political act *but the ultimate instrument of politics.*"[24]

On this point Lenin himself is quite specific. He writes:

> "WAR IS POLITICS BY OTHER
> (i.e. FORCIBLE) MEANS."
>
> This famous dictum belongs to one of the profoundest writers on military questions, Clausewitz. Rightly, the Marxists have always considered this axiom as the theoretical foundation for their understanding of the meaning of every war. It is from this standpoint that Marx and Engels regarded wars.[25]

This being the case, it follows quite naturally to the dialectician that politics is war by other, (i.e., non-forcible) means. Thus Clausewitz has been one of the influences in the development of the Communist doctrine of continuous conflict with the noncom-

23. *Ibid.*
24. *Ibid.*, p. 20.
25. V. I. Lenin, "The Imperialist War: The Struggle Against Social-Chauvinism and Social-Pacifism, 1914–1915," *Collected Works,* Vol. XVIII (New York: International Publishers, 1930), p. 224. Lenin's capitals.

munist world. Mao Tse-tung's concepts of war and revolution also reveal Clausewitz's influence. Mao holds that "politics is war without bloodshed, and war is politics with bloodshed."[26]

Lenin also extracted statements to the effect that, in addition to the enemy military force and territory, the enemy's will to fight is the principal objective of war. He made many notes on elements which Clausewitz generally defines as the "moral factor": morale, leadership, hatred of the enemy, courage, prudence, cowardice, tenacity. He adopted the idea that a defeated country must place its hope in moral superiority and courage, marginally noting "the right to insurrection of the defeated."[27]

Clausewitz writes in *On War* that the aggressor always pretends to be peace-loving because he would like to achieve his conquests without bloodshed and that, therefore, aggression must be presented as a defensive reaction by the aggressor nation. Lenin regarded this as a good idea, and we have seen its use by the Communists in their naked aggressions against Finland, South Korea, Hungary, Laos, Viet Nam and elsewhere. This idea has been and is still a central one in Communist propaganda.

In discussing offensive and defensive warfare, Clausewitz emphasized that an offensive often requires subsidiary defensive action and vice versa. Lenin considered this an excellent example of dialectics, and "active defense" remains a dominant doctrine in Communist thought.

Lenin accepted Clausewitz's idea that there are gradations to war, that it is sometimes more violent, sometimes less. It is probable that the Communist concept of relations with noncommunist states ranging from "not war, not peace" to full-scale war has emerged from this concept.

26. Quoted in Richard L. Walker, *The Continuing Struggle* (New York: Athene Press, 1958), p. 68.
27. Possony, *op. cit.*, p. 21.

Having observed the effects upon the nature of war which had been wrought by the release of new social forces in the French Revolution, Clausewitz had written that each epoch has its own type of war, peculiar to the social and political conditions in which it develops. Of great importance, in his view, was the emergence in his time of the war "of whole nations," made possible by the enthusiasm of the people for the nation's cause. He attributed the French military successes largely to this factor. Lenin agreed, and in his later writings he expresses the view that victory in war derives ultimately from the morale of the forces fighting it.

That Lenin was, in many respects, a disciple of Clausewitz is quite clear from Stalin's later specific disavowal of Clausewitz with the statement that he "has become obsolete as a military authority." In the same statement, however, he stated that "in his reviews of Clausewitz and comments on Clausewitz's book, Lenin did not broach purely military questions of the nature of questions on the military strategy and tactics and their interrelations. . . ." He went on to say that Lenin "praised Clausewitz above all for the fact that the non-Marxist Clausewitz, who enjoyed in his time the authority of an expert on military affairs, supported in his works the well-known Marxist thesis. . . that there exists a direct connection between war and politics, that politics begets war, that war is a continuation of politics by violent means."[28] In view of the fact that Marx wrote long after Clausewitz had died, one can only conclude that Stalin's attempt to credit Marx with Clausewitz's thesis is yet another example of Stalin's effort to deny foreign influence on Soviet thought. In the same paper, Stalin also mentioned his interest, and that of Lenin, in Clausewitz's ideas "on the interrelationship between attack and retreat," "defense and the counteroffensive," and "retreat under certain adverse conditions."[29] The influence of

29. *Ibid.*

28. Josef Stalin, "Comrade Stalin's Answer to a Letter From Comrade Razin," *Bol'shevik* (No. 3: February, 1947), pp. 4–8. Quoted in Raymond L. Garthoff, *Soviet Military Doctrine* (Glencoe, Illinois: Free Press, 1953), p. 55.

Clausewitz on high level Soviet strategic thought could hardly be more apparent.

In terms of the thinking of Soviet military professionals, Clausewitz has also exerted a considerable direct influence. Garthoff describes him as "the most important influence" on Soviet doctrine, and outlines the mechanism of his influence as follows:

> It is. . . very likely that Shaposhnikov, who was a very careful student of Clausewitz' works, introduced much of his thought into Soviet doctrine and into the thinking of many of the leading Soviet marshals of today. Shaposhnikov's work on *The Brain of the Army* (the General Staff) shows his indebtedness to and respect for Clausewitz. Two of the three volumes open with quotations from Clausewitz' writings. Shaposhnikov himself spoke of Clausewitz as "that great philosopher of War." Barmine and other former high Soviet officers attest that in the 1930's Clausewitz' *On War* was being taught.[30]

Clausewitz's influence was also, of course, very strong in the German military staff schools, which were attended by a number of Soviet officers before World War II.

The influence of Clausewitz on the whole spectrum of Soviet military and conflict doctrine is thus very significant, and we shall understand that doctrine better if we first understand Clausewitz. On the other hand, communist conflict doctrine is compounded of many elements, and one should not make too much of the influence of any single one of them. Nevertheless, the core of communist conflict doctrine is Clausewitzian, in precisely the sense that the core of Communist philosophy is Hegelian. Marx and Engels stood Hegel "on his head"; Lenin has stood Clausewitz on his head. Where Hegel held that ideas determine political and economic

30. *Ibid.*, p. 53.

conditions, Marx held that economic conditions determine ideas, and hence determine political institutions. Where Clausewitz held that war is a continuation of politics by other means, Lenin and his successors have substituted the antithetical idea that politics is a continuation of war by other means. The central idea of communist conflict doctrine is thus Clausewitzian, but it is Clausewitz upside down.

Essence of Clausewitz's Strategic Thought

The example set forth above, of the alteration of Clausewitz's concepts to fit an existing pattern of thought, could be multiplied many times by citing examples from German and European uses of Clausewitz's works. As we have seen, the combining of a philosophical and a practical approach by Clausewitz often lends itself to misinterpretation. He has frequently been misunderstood and misused. For that reason, a few clarifications may be useful before the uninitiated reader proceeds to the text.

No System of Warfare Advocated

It is necessary to understand at the outset that Clausewitz did not advocate any particular system of warfare. He went to considerable lengths, in criticizing earlier theories, to show that no universally valid system exists. However, he holds that a theory of war is possible, provided we do not attempt to make it a body of positive rules, a guide to action. Theory should guide the future leader in his self-instruction, "but not accompany him to the field of battle."[31] The necessary knowledge for high military position can be gained only by "experience of life" and "study and reflection."[32] This knowledge must be so thoroughly assimilated into the mind and life of the

31. *See text,* p. 158.
32. *Text,* p. 164.

military leader that it becomes a permanent part of him, enabling him "anywhere and at any moment, to produce from within himself the decision required."[33] Thus, the ability for high military position is to be acquired by study, reflection, and experience. There are no fixed rules, no universal principles, which provide a short cut to success in military life, any more so than in any other field of knowledge.

This aspect of Clausewitz's thought has been frequently, if not consistently, ignored by some of his strongest followers. Many of them have tried to reduce his teachings to simple rules and principles, or have emphasized certain of his concepts in such a way as to give them the standing of a universally applicable principle and to distort the meaning of the whole. A typical example is German emphasis, through World War II, on Clausewitz's concept of the decisive effect of the great battle which destroys the enemy's military forces. This emphasis, it is said, "kept alive the conception of 'true war' within the Prussian officer corps."[34] While it is doubtless true that the emphasis achieved the desirable objective of maintaining an offensive spirit, it also led to the neglect of the role of the defensive and to lack of sophistication and breadth in German military thought. These deficiencies in turn established the climate of military opinion which made it possible for Germany's military leaders to support Hitler because he embodied the offensive spirit, and for Germany to undertake an offensive war when political and economic trends greatly favored a continued and rapid improvement in Germany's power position.

33. *Text,* p. 165.

34. Count Schlieffen, in Introduction to the 5th German edition of *On War* (1905), quoted in H. Rothfels, "Clausewitz," *Makers of Modern Strategy,* ed. by Edward Mead Earle, Gordon A. Craig and Felix Gilbert (Princeton: Princeton University Press, 1943), p. 93.

The Nature of War

Perhaps the most creative part of Clausewitz's work is that which is concerned with examining war to isolate its essence. He defines war as "an act of force to compel our adversary to do our will."[35] Theoretically, the use of force is without limits, but in actual war this is almost invariably moderated by a number of influences. There are thus two kinds of war: one in which each of the adversaries mobilizes all of its means and resources in an effort to overthrow its enemy completely and to avoid being overthrown; the other a war of more limited aim, in which the effort of each of the adversaries is proportionate to the worth of the objective and thus less than total. Every war is a reflection of its own era and the circumstances which bring it about. The greater the tensions and hostility which precede it, the more likely it is to approach its theoretical nature as an act of violence without limit. Generally, however, this image of war must hover in the background as only a possibility, for war is not in real life an independent thing-in-itself but only a part of political relations between national states. It therefore does not make its own laws, but is instead subject to political guidance. It is, therefore, "nothing but a continuation of political intercourse with an admixture of other means."[36]

Clausewitz does not suggest that absolute war cannot occur. In fact, he held that the participation of the whole people had resulted in war in his own time approaching more nearly "its real nature"; that is, an act of violence limited only by the means at the disposal of the participants. He was unwilling to predict that this would always be the case in the future, but he obviously considered it probable.[37] He apparently believed that the tendency toward pure violence is almost a direct function of the extent of mass support for war. This is an interesting thought for our times, since it would

35. See *text*, p. 63.
36. *Text*, p. 255.
37. *Text*, pp. 230–31.

indicate that—at least in a democracy—the difficulties involved in "managing" a war, in the sense of keeping to proper political aims, are very great. No war can be successfully conducted without a reasonable measure of popular support, but a too greatly aroused public opinion in a democracy diminishes the political leadership's freedom of action and increases the tendency for war to become in reality what Clausewitz conceived wars to be only in theory: "struggles of life or death, from pure hatred."

The nature of war is conditioned by its aims. Clausewitz holds that the greater the aim, the more the war will tend toward its absolute conception as an act of pure violence. If the aim of either party is the destruction of the enemy's armed forces in a great battle, the other has no choice but to seek the same end.[38] Each of the opposing states must therefore increase the level of its war effort, in order not to be overthrown and in order to increase the probability of overthrowing the enemy.

For Clausewitz, the destruction of the enemy's armed forces is generally the proper military object of the war. "The destruction of the enemy's forces thus always appears as the superior and more effectual means, to which all others must give way."[39] This destruction need not be confined to the physical, but applies with equal force to the morale of the enemy, his will to resist. The means to the physical destruction is the decisive engagement, the great battle; everything is subject to the decision by arms.

This view has frequently been used to support the assertion that Clausewitz was the "apostle of violence." It will be seen in the text, however, that Clausewitz devotes several chapters to discussion of defensive wars and offensive wars with limited aims, and that his whole conception of what war ought to be—that is, a continuation of state policy with an admixture of other means—is an argument against the folly of war as an act of unrestrained violence. Moreover,

38. *Text,* p. 108ff., esp. p. 112.
39. *Text,* p. 107.

he shows the relationship of ends and means in several discussions, and points out the need to limit the effort to that required to achieve the purpose of the war.[40] When he asserts that if the enemy aims at your destruction, you must then aim at his, he is not advocating a line of action but simply stating what he conceives to be an obvious fact. When he argues that war is, or ought to be, a continuation of state policy by other means, he is clearly arguing against the concept that there is any single object which is suitable for all wars. On the contrary, he suggests that the object of the war must be the success of the policy which led to the war, and that this policy must not be abandoned unless circumstances require it.

Moreover, there is no single way to the object in war: "the defeat of the enemy is not always necessary."[41] If, however, the enemy elects the decision by arms, one cannot refuse. The decisive engagement thus becomes the "supreme law" in war, when it is invoked. It is for this reason that "possible engagements, on account of their consequences, are to be regarded as real ones."[42]

In essence, Clausewitz argues that warfare should be constructive, in the sense that a victory in war should be a victory for the policy that produced the war. On the other hand, Clausewitz asserts that policy must adjust itself to the military means at its disposal. This is an idea which is somewhat more profound than it appears at first glance to be, and it has several implications. First, a policy in peacetime is sound only to the extent that it can be supported militarily if the need should arise. One finds many modern examples of this concept, and it has been most lucidly expressed by General Walter Bedell Smith in his assertion that "diplomacy has rarely been able to gain at the conference table what cannot be gained or held on the battlefield."[43] Second, the objective of a war must be rea-

40. *Text,* pp. 99, 200ff.
41. *Text,* p.99.
42. *Text,* p. 174.
43. Quoted in *Time* (May 5, 1961), p. 22.

sonably attainable with the military means available; otherwise, the objective must be adjusted to accord with the available means, or the necessary means must be created if time and resources permit. Third, the creation of the military means to enforce foreign policy is an essential element of a wise and forward-looking foreign policy. It follows from the preceding three points that mistakes in military policy, during the phase in which weapons systems are being designed and procured and forces of a particular kind created, must later be paid for in adjustments in foreign policy or, if war should occur, in adjustments in the objective of the war. A versatile and flexible military instrument will permit a flexible foreign policy, capable of meeting and adjusting to a wide variety of situations; an inflexible military instrument will necessarily lead to an inflexible foreign policy, extending no further than the means to enforce it.

Strategy

Clausewitz's definition of strategy and tactics has become classic. He asserts that "tactics teaches *the use of the armed forces in engagements,* and strategy *the use of engagements to attain the object of the war.*"[44]

In spite of his insistence, previously cited, on the decisive effect of the great victory, Clausewitz held that the strategic defensive is, in the final analysis, the stronger form of war. This applies, however, only to the defensive rightly understood. The defensive cannot be a mere passive waiting or warding off the enemy's attacks, but must rather be active and vigorous and have as its ultimate aim the acquisition of strength and creation of an opportunity for a decisive attack.

An interesting related idea is Clausewitz's concept of the culminating point in the attack or offensive, beyond which little more can be accomplished and great risks are run.[45] By implication, this

44. *Text,* p. 141.
45. *Text,* p. 269ff.

is the moment for which the defender has been, or should have been, waiting.

The strength of the defensive, moreover, "lies in the nature of all defense; it applies in other spheres of life as well as in war."[46] Clausewitz appears to have in mind here the difficulty of dislodging the possessor or altering the status quo. He asserts that the defense "reaps where it has not sown."

In wars against coalitions, Clausewitz holds that it is usually possible to defeat the coalition by defeating its key member. If this cannot be done or contemplated, the alternative is to regard the war as two or more separate wars.[47]

Beyond these rather general ideas, Clausewitz produced no particular system of warfare, and in fact held that the diversity of the nature and circumstances of war made it impossible to reduce strategy to a set of principles.

Moral and Psychological Factors

In spite of Clausewitz's argument that war should be an instrument of policy, which is to say that it should be rational, he strongly emphasizes the importance of the nonrational human qualities. Several chapters, and many scattered references, on the moral and psychological qualities will be found in the text. Clausewitz justifies his return to the subject by asserting that "the moral elements are among the most important in war."[48] Although recognizing that these elements are hard to define, he nevertheless argues that they belong as much to the theory of war as everything else which constitutes war. His argument for the importance of the moral factor is summed up in his statement that the armies of Europe are generally equal with respect to equipment, training, and discipline. The human factor is therefore the chief element which can change their

46. *Text,* pp. 190–91.
47. *Text,* p. 236.
48. *Text,* p. 177.

relative effectiveness, and this factor is chiefly expressed in the "influence of the national spirit and the habituation of an army to war."[49] In his view, an army becomes habituated to war through war itself or, in peacetime, through extensive realistic maneuvers.

The strength of Clausewitz's views on the moral and psychological factors is perhaps best illustrated by his ringing essay on the unconquerability of a people who will not be conquered. This previously cited essay, called "I Believe and Profess" and written in condemnation of the France-Prussian Treaty of 1812, also contains tersely stated views which were later expanded in *On War.* So far as I can determine, the first part of this essay has not appeared in English, although excerpts from the latter part appear in *On War* and have been included in this edition in the chapter on "Military Virtue of An Army."[50]

In view of the timelessness of "I Believe and Profess," and the insight which it gives into moral strength as the ultimate determinant of the fate of nations, the first part of it has been included in the final section of this edition. In this essay the reader will find a much different Clausewitz from the rational philosopher of *On War.* This is a Clausewitz who is arguing that his country is unconquerable, in the face of the obvious fact that it has been conquered and made subject to Napoleon's will. His argument makes little attempt to appeal to reason; it is a call to the German will. Perhaps no better commentary on the importance of the moral element in war can be made than to recall that this nonrational argument—put forward by Clausewitz and others—eventually enabled the Prussian nation to rise against France at the favorable moment and ultimately to defeat Napoleon completely. This experience is a lesson for subjugated and threatened nations of all times and places, and a confirmation of Clausewitz's views on the importance of the moral and psychological factors. Weapons systems and the conditions and circumstances of

49. *Text,* pp. 180–181.
50. *Text,* pp. 182–88.

war can and do change, but this vital human factor—the determination to resist and to win—remains constantly operative. The words with which Clausewitz ends the emotional argument in the first section of "I Believe and Profess" are therefore as appropriate now as they were in 1812: "the time is yours; what its fulfillment will be, depends upon you. . . ."

A Modern Critique of Clausewitz

Perhaps the greatest of Clausewitz's insights is his recognition that each age—and, within each age, each peculiar condition—produces its own kind of war.[51] Our own era and condition have two characteristic varieties: thermonuclear war and that diffuse, subliminal type which we may term Communist political-guerilla warfare. The first is unique in human experience; the second differs markedly from previous wars of the same species. It is because these kinds of war contain so much that is new or different that one must apply Clausewitz's concepts to them with caution.

Thermonuclear war under present technological conditions contains three new elements. First, its destructive potential is many times that of any previous war, or of all previous wars combined. Second, the delivery of this destructive power by ballistic and air-breathing missiles and by jet aircraft compresses this unprecedented destructiveness into a time period unprecedently short. Finally, there is at present no effective defense against ballistic missiles, and little reason to suppose that defenses against other delivery systems can prevent their weapons reaching a large number of targets. A possibility—which Clausewitz specifically rejected as impossible and therefore excluded from his calculations—thus arises, viz.: that war—thermonuclear war—may now "consist of one blow without duration."[52] It is likely, of course, that some form of military activ-

51. See *text,* p. 204ff., esp. p. 231.
52. See *text,* p. 67ff.

ity would continue after the initial thermonuclear phase, but it also seems likely that the outcome of the thermonuclear phase would largely determine the ultimate outcome of the war.

Under these circumstances, it is quite clear that Clausewitz's concept of the superiority of the defense does not apply: the destructiveness of the attack is overwhelmingly disproportionate to the effectiveness of the defense. Further, Clausewitz argues that "if the issue in war actually depended on single decision or several simultaneous decisions, the preparation for it would naturally have to be carried to the last extreme."[53] He then argues that this cannot occur because all means for the conflict cannot be brought simultaneously to bear. This view is, of course, invalid under present conditions simply because the attacker *can* bring his thermonuclear power to bear simultaneously against the elements Clausewitz listed: the enemy military force, the enemy's country "with its population," and his allies. It is precisely for this reason that preparations for thermonuclear war *are* "carried to the last extreme" of maintaining Western thermonuclear war forces in a posture of readiness that has no peacetime precedent, in order to assure retaliation against such an attack.

This posture is necessary because Clausewitz's concept of the decisive effect of the great battle applies with unique force to the thermonuclear battle. However, the unprecedented destructiveness of thermonuclear war greatly increases the risks, which Clausewitz recognized and cited, of the decisive battle. The thermonuclear battle risks the life of the nation and, since the aim of war must rationally be to preserve the national life, thus introduces a powerful motive to avoid thermonuclear war. This motivation is increased by the fact that the power which prevails in such a war is likely to be so weakened, by the destruction and casualities it suffers and by exhaustion of its store of nuclear and thermonuclear weapons, as to emerge in a position inferior to that of a third power or bloc which has avoided involvement in the war. Finally, we must add the point—which is

53. *Text,* p. 68.

really the first consideration—that, thermonuclear war being morally repugnant, the United States and its allies have rejected any resort to it except under circumstances which appear to offer no other choice. Thus, rationally and morally, the great decision by thermonuclear battle is not an acceptable means for "the continuation of policy" for the Western world. We have elected to employ other measures for our defense in that struggle which is the central feature of our age.

If we could be permanently certain of a Communist commitment to do likewise, we should be able to reduce and perhaps ultimately to forego thermonuclear power. It is here, however, that Clausewitz's concept of the great decision by arms still sharply applies, for it is still true—as Clausewitz asserted—that if the enemy aims at our destruction, we must then aim at his. Failure to comprehend this simple proposition is the source of error in the now rejected "minimum deterrence" concept. The effective and credible capability to fight the great thermonuclear battle, if we must, is the most effective—and in some cases probably the only—means to underwrite any policy or action which risks war with a major thermonuclear power. Put another way, strategic thermonuclear delivery capability is the central, dominant element in modern power relationships; it is the fulcrum upon which all else rests, and cold war actions or conflict measures short of war are, like levers, dependent upon the adequacy of that fulcrum, as well as upon their own adequacy, for their effectiveness.

For nonnuclear war or for any conflict between a nuclear power and a nonnuclear power, the concept of the decisive effect of the great victory is, of course, still clearly valid. Here, however, all the important reservations which Clausewitz attached to the concept apply.

Guerilla warfare represents a special case for the concept of the decisive effect of the great battle, for it is precisely the ability of guerilla forces to avoid a decisive engagement which gives them their viability. By being diffused and scattered, yet stronger than any forces in the country except the regular force opposing them, they can control much of the country and yet avoid battle by simply melting away when regular forces enter their areas. In guerilla warfare supported from outside the country, a few men can grow to a squad, a squad to a platoon, a platoon to a company, a company to a battalion—in short, the diffused guerilla bands can become an army. When this metamorphosis has occurred, the guerilla army can receive heavy equipment and then elect or avoid the decisive engagement, as it chooses. The essential point is that only the guerilla force can elect the great decision by arms—it is not ordinarily a course of action which can be elected by their regular army opponents.

It follows, from the above discussion, that the great decisive engagement is unlikely, under present conditions, to appear as a feasible course of action for the Western world in either general war or counter-guerilla warfare. We must accept the probability of a long struggle, in which final success or failure will be the sum of many small gains or losses.

This probability is not necessarily discouraging for, in terms of the larger problem associated with the defense of the noncommunist world, Clausewitz's doctrine of the inherent superiority of the defense offers some hope that the defense—rightly understood and employed—can lead ultimately to stabilization and, if necessary, to the offensive. The fact that the Western world is currently on the defensive does not govern the eventual outcome of the struggle between communism and representative government. The defensive, moreover, need not be a permanent aspect of Western political policy, for the appeals of representative government and free societies are very powerful.

Turning now to Clausewitz's deficiencies, we find—as we might expect—that Clausewitz's thought falls short of the modern dimension in a number of respects. For example, he gave little attention to the intimate interrelationship and mutual influence of technological change and strategic thought. (This factor was, of course, of less importance in his time than it is now, and his thinking was also perhaps colored by his view that the technologies of the

European powers were roughly equal.) He did not weigh economic factors in warfare, in spite of the fact that he had witnessed Napoleon's organization of the "continental system" as a major economic warfare offensive. A similar source of mild wonderment—considering his emphasis on the moral factors in war and his identification of destruction of the enemy's will to fight as a proper aim of military action—is the fact that he produced no systematic discussion of psychological warfare measures.

In all these matters, of course, Clausewitz's writing reflects the general level of the military thought of his time and place. To give him his due, we must add that he was writing under political constraints between 1818 and 1830. It seems certain that his awareness of these nonmilitary factors in war was greater than his writings reflect.

As a soldier, Clausewitz was not an innovator: he did not change warfare *in his time,* nor is the world indebted to him for founding any new system of strategy. His was an inquiring and critical mind, rather than a creative one: he analyzed, understood and codified the methods of warfare which others had created. In doing so, he performed an enormously creative act, for he changed the future by seeing and revealing to others the larger significance which lay below the surface of contemporary events.

Even in this, his thought seems to have led his German disciples more often to disaster than to triumph. For this, Clausewitz is not to blame, for he wrote for men of wider vision than his influential German followers proved to be. His German successors, from the elder Moltke through Schlieffen, Ludendorff, and Hitler, interpreted Clausewitz largely in terms of the preeminence of aggressive war to achieve national aims, learning nothing from the great defeat of World War I, and never seeing the broad canvas of war as a whole.[54] Even as he stood in the ruins of his country in April 1945 and gave his final testament to the German people, Hitler referred

54. See Colonel Vincent J. Esposito, "War As A Continuation of Politics," *Military Affairs,* XVIII (Spring, 1954), p. 21.

to "the great Clausewitz."[55] Hitler's fate, and Germany's, must await all who see war only as a matter of armies, invasions, great victories and much bloodshed. Clausewitz saw and codified these aspects, but he also saw more: he understood that a dialectical relationship exists between war and politics, social strength and military strength, national policy and military means, offense and defense, change and status quo. He saw that *war is a subject for scholarly study and inquiry—a fact which we are only beginning to grasp—and that this study must go far beyond the purely military aspects. In recognizing that there could be a critical philosophical approach to the study of war, he anticipated the growing modern recognition that the study of war is as much a matter for the politician, the diplomat, the scientist and the professor as for the professional officer, if they will but come to the soldier's aid. This is true because there is much in war that is not military.* The modern military professional has grasped this fact, and sought with some success to extend his horizons to the political, economic, technological and other aspects of war. This effort can never be a complete success; the complexity and diversity of modern war, and of conflict short of war, are too great, and the military profession itself too demanding, for the representatives of that profession alone to carry forward the study of war in all its aspects. *What is needed is systematic study of war by competent members of all the disciplines involved in war.*

Thus, in answer to the question propounded by Clausewitz's critics, "qu'a-t-il fondé?" our answer must be: Clausewitz founded the modern basis for the systematic study of war as a field of human knowledge. That, in doing so, he was far ahead of his time is attested by the fact that the world has long studied *war-making,* but is only now beginning the serious study of *war* as social, psychological, political, economic, and technological phenomenon. He has shown us the road; by following it until we reach an understanding of war, we may yet understand how to have peace.

55. *Ibid.*

I

On War

I

WHAT IS WAR?

WE PROPOSE TO PROCEED from the simple to the complex, covering first the several elements of our subject, then its parts or divisions, and finally the whole. It is first necessary, however, to glance at the nature of the whole, because in this subject, more than any other, the whole and the part must be considered together.

We must first define war. We shall not begin with a pedantic definition, but confine ourselves to war's essence: the duel. War is nothing but a duel on a larger scale. If we would unite in one conception the countless duels of which it consists, we should imagine two wrestlers. Each seeks by physical force to overthrow the other, render him incapable of further resistance, and compel his opponent to do his will.

War is thus an act of force to compel our adversary to do our will.

. . . Physical force is the *means;* to impose our will on the enemy is the *object*. International law and usage impose minor restrictions, which do not really weaken its power, on the use of force. To achieve the object with certainty requires that the enemy be disarmed, and this disarming is by definition the proper aim of military action. In a sense, it thrusts the object aside and takes its place.

The use of force is theoretically without limits. Philanthropic souls may imagine that there is a way to disarm or overthrow our adversary without much bloodshed, and that this is what the art of war should seek to achieve. Agreeable as it may sound, this is a false idea which must be demolished. As the greatest use of force does not exclude the cooperation of intelligence, the ruthless user of force who shrinks from no amount of bloodshed must gain an advantage if his opponent does not do the same. Thus each drives the other to extremes which are limited only by the adversary's strength of resistance.

War is nothing but a duel on a larger scale.

The matter must be regarded in this light; it is a waste of effort to ignore the element of brutality because of its repugnance.

If the wars of civilized nations are far less cruel and destructive than those of the uncivilized, the cause lies in the social conditions of these states, internally and in their relations with each other. These conditions shape, limit and modify war, but they do not themselves belong to war: they already exist. We can never introduce a modifying principle into the philosophy of war without committing an absurdity. . . .

Although savages are inspired more by emotion and civilized peoples more by intelligence, even the most civilized nations can be passionately inflamed against each other. We should therefore miss the truth if we ascribed war among civilized men to a purely rational act of governments and conceived it as continuously reducing the element of passion. . . .

Theory was beginning to move in this direction when the events of the war with Napoleon taught us better. War is an act of force, and the emotions are necessarily involved in it. If war does not originate from them, it more or less reacts upon them, and the degree of this is not dependent upon the stage of civilization but upon the importance and duration of the hostile interests.

So we repeat: War is an act of force, and there is no limit to the application of that force. Each of the adversaries forces the hand of the other, and a reciprocal action results which theoretically can have no limit. This is the first reciprocal action that we meet and the first extreme.

We have said that the aim of military action is to disarm the enemy. We shall now show that, at least theoretically, this is necessarily so.

If our opponent is to do our will, we must put him in a position more disadvantageous to him than the sacrifice which we demand of him. The disadvantage must not appear transitory, or he would wait for a more favorable time and refuse to yield. Continuance of military activity must thus appear to him to lead to a less advantageous position. Since the worst situation for him is that of being completely disarmed, we must either actually disarm him or create a condition that threatens him with the probability we will do so. It follows, then, that the disarming or the overthrow of the enemy—whichever we choose to call it—must always be the aim of military action.

Now war is always the collision of two live forces with each other, and this aim must be assumed to apply to both sides. So long as I have not overthrown my enemy, I must beware that he may overthrow me. He forces my hand as I force his. This is the second reciprocal action, which leads to the second extreme.

War tends toward the utmost exertion of forces. If we wish to overthrow our opponent, we must proportion our effort to his power of resistance. This power is a product of two factors: *the extent of the means at his disposal and the strength of his will.* It is possible to estimate the extent of the means at his disposal, but the strength of his will can only be approximated by the strength of the motive behind it. Having obtained a reasonably probable estimate of our opponent's power of resistance, we can proportion our efforts so as to reach a preponderance or, if our means do not suffice for this, do as much as we can. But our opponent does the same, and a fresh competition arises which, in pure theory, once more involves pushing to an extreme. This is the third reciprocal action.

The theoretical tendency toward the utmost in exertion is modified in practice. . . . If we wanted to deduce, from a mere theoretical conception of war, an absolute aim and the means to be employed to achieve it, continuous reciprocal actions would lead us to fanciful extremes. If, adhering to an absolute aim, we proposed to insist that we must always be prepared for the extreme of resistance and meet it with the extreme of effort, we should be insisting on a mere paper law with no application to the real world.

. . .However, everything assumes a different shape if we pass from this abstract world to the real. In the abstract we had to conceive both one side and the other as not merely striving for perfection but achieving it. Will this ever be so in practice? It would if:

1. War were a wholly isolated act, which arose suddenly and without connection with the previous course of events.
2. It consisted of a single decision or of several simultaneous decisions.
3. Its decision were complete in itself and the ensuing political situation were not already being taken into account and influencing it.

With reference to the first of these three points, war is never an isolated act. In the real world, war never breaks out suddenly, and it does not spread immediately. Each of the opponents can thus form an opinion of the other from what he is and does, not from what he theoretically should be and do. Moreover, human organizations always fall short of the absolute best, and these deficiencies, operative on both sides, become a modifying influence.

With regard to the second of the three points, war does not consist of one blow without duration. If the issue in war actually depended on a single decision or several simultaneous decisions, the preparation for it would naturally have to be carried to the last extreme. In the real world, however, the decision consists of several successive acts, each influencing those which follow, and thus the trend to the extreme is modified.

Every war, however, would necessarily be confined to a single decision or several simultaneous decisions if all means for the conflict were brought into operation simultaneously or could be so brought into operation, because an adverse decision necessarily diminishes these means. If they have all been used up in the first decision, a second becomes unthinkable. All acts of war which could follow would be essentially part of the first and really only constitute its duration.

. . .However, the very nature of these resources makes it impossible to put them all into operation simultaneously. They consist of *the military forces proper, the country with its population,* and *the allies.*

. . .All movable military resources can be put into operation simultaneously, but not the whole country, unless it is so small as to be wholly embraced by the first act of war. Moreover, the cooperation of the allies is not subject to the will of the belligerents, and—because of the very nature of political relations—it frequently does not come into play until later, to restore a balance of forces which has been upset.

. . .It is thus contrary to the nature of war to make all our resources available at once. But, what either opponent omits from weakness becomes, for the other, an objective ground for relaxing his efforts. Through this reciprocal action, the trend toward the extreme is once more reduced to a limited effort.

Lastly, the final decision of a whole war is not always an absolute one. The defeated state often sees it as a transitory evil, for which a remedy can yet be found in later political circumstances. How greatly this must also modify the violence and the intensity of effort is obvious.

In this way the probabilities of real life replace the extremes of theory. The whole field of war ceases to be subject to the strict law of forces pushed to the extreme. In the real world, it is left to judgment to determine the limits of effort, applying the laws of probability to data supplied by real world phenomena. The actual situation supplies the data for judging what is to be expected. Each side will draw its conclusions from the character, the institutions, the situation and the circumstances of the enemy; each will judge, according to the laws of probability, what the other's action will be and will govern itself accordingly.

As the law of the extreme—the intention of disarming and overthrowing the enemy—loses its force, the political object of the war once more comes to the front. The political object as the original motive of the war will be the standard for both the aim to be attained by military action and for the efforts required for this purpose. It cannot be in itself an absolute standard but, as we are dealing with reality and not with mere ideas, it will be the standard relative to the two contending states. One and the same political object can in different nations, and even in the same nation at different times, produce different reactions. The political object can therefore be allowed to serve as a standard only in so far as we bear in mind its influence on the masses which it is to affect. Accordingly, the character of these masses must be considered. The action may be strengthened or weakened by the feeling of the masses, with quite different results. The tensions and hostile feelings between states may be so great that a very trifling motive for war can produce a wholly disproportionate effect—a positive explosion.

It is thus contrary to the nature of war to make all our resources available at once.

This holds good for the efforts which the political object can evoke in the two states, and for the aim it can assign to military action. Sometimes the political object can itself become this aim—for example, if it is the conquest of a certain province. Sometimes the political object will not be suited to provide the military aim; in such cases an aim must be chosen which can serve as an equivalent for the political object and take its place in the conclusion of peace. In this case, however, due consideration for the character of the states concerned is always essential. There are circumstances in which the equivalent must be much greater than the political object, if the latter is to be attained by it.

The greater the indifference of the masses, and the less serious the tensions which exist on other grounds between the states, the more dominant and decisive will the political object be. There are cases in which it is, almost by itself, the deciding factor.

Now, if the aim of the military action is an equivalent for the political object, that action will in general diminish as the political object diminishes. The more the political object comes to the front, the more this will be so. Thus there can be wars of all degrees of importance and energy, from a war of extermination down to a mere state of armed observation. This leads us now to another kind of question which must be analyzed and answered.

The question is: Can military action in war be suspended even for a moment, regardless of how insignificant the political claims of either side or how weak the means and how trivial the military aim? This question goes deep into the essence of the matter.

The accomplishment of every action requires a certain time, which we call its duration. . . . If we allow every action in war its duration, we must admit that every expenditure of time in excess of this—that is, every suspension of military action—seems at first sight to be absurd. However, the question is not of the progress of one or the other of the opponents, but of the progress of the military action as a whole.

It seems that only one cause can suspend action, and that this seems possible on only one side. So long as two warring parties remain under arms and do not make peace, the original hostile motive which caused the conflict must remain, and this motive can cease to act upon either of the two opponents for only one reason: *that he wants to await a more favorable moment for action.* It is obvious that this reason can be present on only one side, because if it is to the interest of one commander to act, it must be to the interest of the other to wait.

A complete equilibrium of forces can never produce a suspension of action, for in such a suspension the assailant, who has the positive aim, must necessarily retain the initiative.

However, if we conceive the equilibrium as one in which he who has the positive aim, and therefore the stronger motive to act, has the smaller forces available—so that the equation would arise from the product of motives and forces—we should still have to say that if no change is foreseeable in this equilibrium, both sides must make peace. If a change is to be foreseen, it will be in favor of one side only, and for that reason the other will be constrained to act. The idea of an equilibrium thus cannot explain a suspension of action; all it amounts to is the awaiting of a more favorable moment. Let us assume, therefore, that of two states one has the positive aim; for example, the conquest of one of the opponent's provinces to be used as a counter in the peace settlement. His political object is attained by this conquest; his need for action ceases and he can rest. If his adversary is prepared to accept this result, he must make peace; if not, he must act. But, if he considers that he will be better prepared to act in four weeks' time, he has sufficient grounds for postponing action.

From that moment the duty to act seems to fall logically upon his opponent, so that no time may be allowed the vanquished to prepare for action. It is assumed, of course, that in all this each side has a complete knowledge of the circumstances.

In this way a continuity of military action would be introduced, forcing everything again to a climax. If this continuity of military action actually existed, everything would again be forced to the extreme. . . . We know that military action seldom or never has this continuity, and that there have been many wars in which action occupies by far the smaller part of all the time expended, and inaction all the rest. Suspension of action must therefore be possible, and not a contradiction in itself. The fact that this is so, and the reasons for it, we will now show.

By supposing the interests of one commander to be always diametrically opposed to those of the other, we have assumed a true polarity. . . .

However, attack and defense are unlike things, and of unequal force. Polarity therefore is inapplicable to them. If war had only one form, the attack, and therefore no defense; if, in other words, only the positive motive distinguished the attack from the defense and the methods of the fight were always the same, every advantage to one side would be an equal disadvantage to the other and a true polarity would exist.

But military action has two separate forms, attack and defense, which are very different and of unequal strength. Polarity therefore lies in the relation of attack and defense to the decision to attack or defend, but not in attack or defense *per se*. If one commander wishes to postpone a decision, the other must wish to hasten it. If it is to A's interest not to attack at once but four weeks hence, it is to B's interest to be attacked by A at once and not four weeks hence. Here is a direct opposition of interest, but it does not follow from it that it is to B's interest to attack A at once. That is obviously something quite different.

The effect of polarity is often destroyed by superiority of the defense to the attack. This explains the suspension of military action. If the defensive form in war is, as we shall show, stronger than the offensive form, the question arises as to whether a deferred decision to attack is as advantageous for one side as the defense is for the other. If the side for which the present is favorable is too weak to dispense with the advantages of the defensive, it must accept a less favorable future. It may still be better to fight a defensive battle in the less favorable future than an offensive one in the present, or than to make peace. Now, as we are convinced that the superiority of the defense (rightly understood) is very great, a very large proportion of the periods of inaction which occur in war are thereby explained, without our being necessarily involved in a contradiction. The weaker the motives to action are, the more they will be neutralized by this difference between attack and defense. The more frequently, therefore, will military action be suspended, as indeed, experience teaches.

Imperfect knowledge of the situation is another cause for suspension of military action. No commander has accurate knowledge of any position but his own; his adversary's is known to him only through uncertain reports. Through mistakes in his judgment of these reports, a commander can believe that the initiative lies with his opponent when it really lies with himself. Although it is true that this lack of knowledge could just as often cause untimely action as untimely inaction, and would in itself no more contribute to delay than to hasten military action, it must still be regarded as one of the natural causes that, without involving an internal contradiction, may halt military action. If we reflect upon how much more we are inclined to overestimate rather than underestimate the strength of our opponent, because it lies in human nature to do so, we must admit that imperfect knowledge of the situation must generally contribute greatly to stopping military action and modifying the principles upon which it is conducted.

The possibility of a standstill introduces into military action a new modification by diluting it with the element of time and increasing the means of restoring a lost balance of forces. The greater the tensions out of which the war has sprung, the shorter will be these periods of inaction; the weaker the hostile feeling, the longer they will be. Stronger motives increase the power of the will which, as we know, is always a factor in the product of our forces.

Frequent periods of inaction remove war still further from the realm of exact theory and make it still more a calculation of probabilities. The more slowly military action proceeds and the longer and more frequent the periods of inaction, the more readily a mistake can be repaired, and thus the bolder the commander will become in his assumptions, and at the same time the more he will tend to remain below the extreme demanded by theory and build everything upon probability and conjecture. The more or less leisurely course of military action accordingly allows more or less time for what the nature of the concrete situation already demands, a calculation of probabilities in accordance with the given circumstances.

The foregoing shows how much the objective nature of war makes it a calculation of probabilities. It now needs but one additional element to make it a gamble, and that element—the element of chance—it certainly does not lack. There is no other human activity that stands in such constant and universal contact with chance as does war. Thus chance, the accidental, and good luck play a great part in war.

If we now glance at the subjective nature of war—that is, at those qualities with which it must be carried on—it must strike us as still more like a gamble. The element in which the activity of war moves is danger. In danger, courage is the highest of all moral qualities. Now, courage is certainly quite compatible with prudent calculation, but courage and calculation are nevertheless different in kind and belong to different parts of the mind. On the other hand, daring, reliance on good fortune, boldness and foolhardiness are only manifestations of courage, and all these efforts of the spirit seek the accidental because it is their proper element.

We thus see that the absolute, the so-called theoretical, faculty finds nowhere a sure basis in the calculations of the art of war. From the outset there is a play of possibilities and probabilities, of good luck and bad, which permeates every aspect of war, great or small, and makes war, of all branches of human activity, the most like a game of cards.

We must now examine how this accords with the human mind. Although our intellect always feels itself urged toward clarity and certainty, our mind still often feels itself attracted by uncertainty. Instead of threading its way with the intellect along the narrow path of philosophical investigation and logical deduction, in order, almost unconsciously, to arrive in strange and unfamiliar territory, it prefers to linger with the imagination in the realm of chance and luck. Instead of being confined, as in the first instance, to meager necessity, it revels here in the wealth of possibilities. Enraptured thereby, courage takes wings, and flings itself into the element of daring and danger as a fearless swimmer flings himself into the stream.

Shall theory leave it here and move on, self-satisfied, to absolute conclusions and rules? In that case theory is of no practical use, for theory must also consider the human element and accord a place to courage, boldness, and even foolhardiness. The art of war has to do with living, moral forces. It therefore follows that it can nowhere attain the absolute and certain; there remains always a margin for the accidental, in great things and small. Courage and self-confidence must fill the gap left by this accidental element. The greater the courage and self-confidence, the larger the margin that may be left to the accidental. Courage and self-confidence are thus principles absolutely essential to war. Theory must therefore lay down only such rules as allow free scope for these necessary and noblest of military virtues in all their degrees and variations. Even in daring there is still wisdom and prudence, only they are estimated by a different standard of values.

There is no other human activity that stands in such constant and universal contact with chance as does war.

Such is the nature of war, such the commander who conducts it, and such the theory that governs it. But war is no pastime, no mere passion for daring and winning, no work of an uninhibited enthusiasm; it is a serious means to a serious end. All that it displays of the glamour of fortune, of the thrills of passion and courage, of imagination and enthusiasm, are only particular properties of this means.

We shall here attempt to define war more particularly.

The war of a community—of whole nations, particularly civilized nations—always arises from a political condition and is called forth by a political motive. It is therefore a political act. Now if war were an act complete in itself, an absolute manifestation of violence, as we had to deduce from its mere conception, it would, from the moment it was initiated by policy, step into the place of policy and, as something quite independent of policy, set it aside and follow its own laws. . . . War has hitherto been regarded in this way even in practice, whenever a lack of harmony between policy and the conduct of war has led to theoretical distinctions of this kind. But it is not so, and this idea is radically false. War in the real world is, as we have seen, no such extreme thing releasing its tension in a single discharge. It is the operation of forces which do not in every case develop in the same way and the same proportion, but which at one moment arise to a sufficient pitch to overcome the resistance which inertia and friction oppose to them, while at another, they are too weak to produce any effect. War is, therefore, a regular pulsation of violence, more or less vehement and consequently more or less quick in relaxing tensions and exhausting forces—in other words, more or less quickly leading to its goal. Nevertheless, it always lasts long enough to influence that goal, so that its direction can be changed; in short, long enough to remain subject to the will of a guiding intelligence. Now, if we reflect that war originates in a political motive, we see that this motive remains the first and highest consideration in its conduct. The political object is not, however, on this account a despotic lawgiver; it must adapt itself to the means at its disposal—and is often completely changed by this adaptation—but it must always be the first thing to be considered. Policy will therefore permeate the whole action of war and exercise a continual influence on it, so far as the nature of the explosive forces in it permit.

War is therefore a continuation of policy by other means. It is not merely a political act but a real political instrument, a continuation of political intercourse, a conduct of political intercourse by other means. What still remains peculiar to war relates merely to the peculiar character of the means it employs. The art of war in general, and the commander in each particular case, can demand that the tendencies and designs of policy shall not be incompatible with these means. The claim is certainly no small one, but regardless of how powerfully it may influence political designs in particular cases, it must still be regarded always as only a modification of them; for the political design is the object, while war is the means, and the means can never be thought of apart from the object.

The nature of wars is diverse. The greater and more powerful the motives for war, the more they affect the whole existence of the nations involved, and the more violent the tensions which precede war, the more closely will war conform to its abstract conception. It will be more closely concerned with the destruction of the enemy—the military and the political aim will more closely coincide—and will tend to be more purely military and less political. The weaker the motives and the tensions, the less will the natural tendency of its military element—the tendency to violence—coincide with the directives of policy. The nature of a war thus tends to be governed by the motives which produce it.

To avoid being misunderstood, we must here remark that by this natural tendency we mean only the philosophical, the strictly logical tendency of war and not the emotions and passions of the forces actually engaged in combat. It is true that in many cases these emotions might be excited to such a pitch that they could be kept confined to the political road only with difficulty, but in most cases such a contradiction will not arise because the existence of such strong emotions will imply also the existence of a great plan in harmony with them. If the plan is directed only to an inconsequential object, the emotional excitement of the masses will be so slight that they will always be more in need of urging than of restraining.

To return to our main subject: All wars may be regarded as political acts. Although it is true that in one kind of war policy seems entirely to disappear, while in another it very definitely comes to the fore, we can nevertheless maintain that the one kind is as political as the other. If we regard policy as the intelligence of the state, we must include, among the situations which policy must weigh, that in which all the circumstances postulate a war of the first kind, in which policy disappears. The latter kind of war could belong to policy more than the former only if we understood the term policy not as a comprehensive knowledge of the situation but as it is conventionally thought of—a cautious, crafty, even dishonest cunning, averse to violence.

We see first, therefore, that we must in all circumstances think of war not as an independent thing, but as a political instrument. Only by adopting this viewpoint can we avoid falling into contradiction with the whole of military history. This alone opens the great book to intelligent appreciation. Secondly, this same viewpoint discloses how wars must differ according to the nature of their motives and the circumstances out of which they arise. In this regard, the greatest and the most decisive act of judgment which a statesman and commander perform is that of recognizing correctly the kind of war in which they are engaged; of not taking it for, or wishing to make of it, something which under the circumstances it cannot be. This is, therefore, the first and most comprehensive of all strategic questions. We shall examine it more closely later on, in the chapter on the plan of a war. For the moment we content ourselves with having established the main point of view from which war and the theory of war must be regarded.

All wars may be regarded as political acts.

What is the result for theory? We see that war is therefore a chameleon, because in each concrete case it changes somewhat its character. It is also, when regarded as a whole consisting of all the tendencies predominating in it, a strange trinity. It is composed of the original violence of its essence, the hate and enmity which are to be regarded as blind, natural impulse; of the play of probabilities and chance, which make it a free activity of the emotions; of the subordinate character of a political instrument, through which it belongs to the province of pure intelligence.

The first of these three sides is more particularly the concern of the people, the second that of the commander and his army, the third that of the government. The passions which flame up in war must be already present in the peoples concerned; the scope that the play of courage will get in the realm of probability and chance depends upon the character of the commander and the army; the political objects are the concern of the government alone.

These three tendencies, these three lawmakers, lie deep in the nature of the subject and at the same time vary in magnitude. A theory which left one of them out of account, or attempted to fix an arbitrary relation between them, would immediately contradict reality and be invalidated.

The problem, therefore, is that of keeping the theory poised between these three tendencies as between three centers of attraction.

We propose to investigate, in the book dealing with the theory of war, how this difficult problem can be solved in the most satisfactory way. . . .

II

ENDS AND MEANS IN WAR

HAVING ASCERTAINED THE COMPLEX and variable nature of war, we shall now consider what influence this has upon the means and the end in war.

If we ask, first of all, what is the aim toward which the whole war must be directed so as to be the proper means for attaining the political object, we shall find that this is just as variable as are the political object and the particular circumstances of the war.

If we begin instead by keeping once more to pure theory, we are bound to say that the political object of war really lies outside war's province, for if war is an act of violence to compel the enemy to do our will, then in every case everything would necessarily and solely depend on overthrowing the enemy; that is, on disarming him. This object, which is decided from pure theory but which in reality is nearly approximated in a large number of cases, we shall first of all examine in the light of this reality. Later on, in the plan of a war, we shall consider more closely what disarming a state means, but we must here distinguish between three general categories which include everything else. They are the *military forces,* the *country* and the *will of the enemy*.

The *military forces* must be destroyed; that is, put into such condition that they can no longer continue the war. In what follows, the expression "destruction of the enemy's military forces" is to be understood only in this sense.

The *country* must be conquered, for out of the country fresh military forces could be raised.

Even if both of these things have been done, the war—that is, the hostile tension and the activity of hostile agencies—cannot be regarded as ended so long as the *will* of the enemy is not subdued; that is, until his government and his allies have been induced to sign a peace or his people to submit. Even though we are in full possession of the enemy's country, the conflict may break out again in the interior or through assistance from his allies. No doubt this may also happen after the peace, but this only shows that wars do not always contain the elements necessary for a complete decision and settlement. Moreover, even when this is the case, the conclusion of a peace always extinguishes sparks which would have gone on smoldering, and the tensions slacken because the minds of those who are inclined toward peace—and in every nation and in all circumstances there are always a large number of these—turn wholly away from the idea of resistance. Thus, whatever may occur subsequently, we must always regard the end as attained, and the business of war as finished, with the peace.

Of the three things enumerated above, the military forces are meant to defend the country. The natural order, then, is that these should first be destroyed, then the land should be conquered, and as a result of these two successes and the strength which we shall still possess, the enemy should be induced to make peace. The destruction of the enemy's military forces usually occurs by degrees, and in a corresponding measure the conquest of his country immediately follows. The two things usually interact, the loss of the provinces helping to weaken the military forces. However, this order is by no means necessary, and for that reason it is not always followed. The enemy forces may, even before they have been noticeably weakened, retreat to the opposite side of the country or even into foreign territory. In this case, the greater part of the country, or even the whole, is conquered.

But the disarming of the enemy—this object of war in the abstract, this final means of attaining the political object, in which all other means are included—does not always occur in practice and is not a necessary condition to peace. Therefore it cannot be set up in theory as a law. There are innumerable instances in which peace treaties have been concluded before either of the parties could be regarded as disarmed; indeed, even before the balance of strength had undergone any noticeable alteration. Moreover, if we look at actual cases, we must admit that in a whole class—the cases in which the enemy is distinctly stronger—the overthrow of the enemy would be a futile playing with ideas.

The reason why the object of war deduced from pure theory is not generally applicable to real war lies in the difference between the two, which was discussed in the preceding chapter. Pure theory would hold that a war between states of very unequal military strength would be an absurdity and therefore impossible. The inequality in physical strength would have to be, at most, no greater than could be offset by moral strength, and that would not go far in our present social condition in Europe. If, therefore, we have seen wars occur between states of very unequal military strength, it is because real war is often far removed from our original theoretical conception of war.

There are two considerations which in practice can take the place of the impossibility of further resistance as motives for making peace. The first is the improbability of success, the second an excessive price to be paid for it.

As we have seen in the previous chapter, war must always free itself from the strict law of logical necessity and rely upon the calculation of probabilities. This is always much more the case when the war is inclined in that direction by the circumstances from which it has sprung; that is, the less powerful its motives and the tensions it has produced. This being so, it is quite conceivable that out of this calculation of probabilities even motives for making peace may arise. A war need not, therefore, always be fought out until one of the parties is overthrown, and we may suppose that when the motives and passions are weak, a scarcely perceptible probability is sufficient to move the side to which it is unfavorable to give way. Now, were the other side convinced of this beforehand, it would naturally strive for this probability only, instead of wasting effort in an attempt to overthrow the enemy completely.

Still more general in its effect on the decision to make peace is the consideration of the expenditure of force already made and further required. As war is no act of blind passion, but is dominated by the political object, the value of that object determines the measure of the sacrifices by which it is to be purchased. This will be the case for both the extent and the duration of the sacrifices. As soon, therefore, as the required expenditure of force exceeds the value of the political object, the object must be abandoned, and peace will be the result.

We see, therefore, that in wars in which the one side cannot completely disarm the other, the motives to peace will rise and fall on both sides according to the probability of success and the required expenditure of force. If these motives were equally strong on both sides, they would meet in the center of their political difference. What they gain in strength on the one side they should lose on the other. If their combined sum is sufficient, peace will result, but naturally to the advantage of the side which has the weakest motives for its conclusion.

At this point we purposely pass over, for the moment, the difference which must necessarily result in practice from the *positive* or *negative* character of the political object. Although this is of the highest importance, we must keep here to a still more general point of view, because the original political intentions change greatly during the course of the war and may ultimately become totally different, *just because they are partly determined by the successes and by the probable results.*

The question of how the probability of success can be influenced now arises. In the first place, naturally, it can be influenced by the same means we use when the object is the overthrow of the enemy; that is, by the destruction of his military forces and the conquest of his provinces, though neither of these means are of quite the same import here as they would be in connection with that object. If we attack the enemy's forces, it is a very different thing whether we intend to follow up the first blow with a succession of others, until the whole force is destroyed, or whether we mean to content ourselves with one victory in order to shatter the enemy's feeling of security, convince him of our superiority, and so instill in him apprehensions about the future. If this is our object, we go only so far in the destruction of his forces as is sufficient for the purpose. In like manner, the conquest of the enemy's provinces is quite a different measure if our object is not the overthrow of the enemy. If this were our object, the destruction of the enemy's forces would be the only really effective action, and the taking of the provinces only the consequence of it. To take them before his forces had been shattered would always have to be regarded as a necessary evil. On the other hand, if our purpose is not the overthrow of the enemy, and if we are convinced that the enemy does not seek but fears to bring matters to a bloody decision, the taking of a weak or undefended province is an advantage in itself. If this advantage is sufficient to make the enemy apprehensive about the final result, then it may also be regarded as a shorter road to peace.

But now we come upon another particular means of influencing the probability of success without defeating the enemy's armed forces, namely, upon enterprises which have a direct influence on policy. If there are any enterprises which are particularly suited to breaking up the enemy's alliances or making them ineffective, to winning new allies for ourselves, to stimulating political activities in our favor, and so forth, then it is easy to understand how much these may increase the probability of success and become a much shorter route to our object than the defeat of the enemy's armed forces.

The second question is how to influence the enemy's expenditure of strength; that is, how to raise for him the price of success. The enemy's expenditure of strength lies in the *wastage* of his forces, consequently in the *destruction* of them on our part, and in the *loss of provinces,* consequently the *conquest* of them by us. Upon closer examination it will be evident that each of these terms varies in meaning and that the operation it designates differs in character according to the object it has in view. The fact that the differences will generally be slight should not cause us perplexity, for when motives are weak, the finest shades of difference are often decisive in favor of this or that method of applying force. Our only concern here is to show that, certain conditions being supposed, other ways of attaining our object are possible, and that they are not contradictory, absurd, or even mistaken.

Besides these two means, there are three other ways of directly increasing the enemy's expenditure of forces. The first is *invasion,* that is, the *occupation of the enemy's territory, not with a view to keeping it,* but in order to levy contributions upon it or devastate it. The immediate object here is neither the conquest of the enemy's territory nor the defeat of his armed forces, but merely to *do him damage in a general way.*

The second way is to direct our enterprises to those points at which we can do the enemy the most harm. Nothing is easier to conceive than two different directions in which our forces may be employed, the first of which is greatly preferable if our object is to defeat the enemy's armed forces, while the other is more advantageous if defeat of the enemy is out of the question. According to the usual manner of speaking, the first would be considered the more military course, the second the more political. From the highest point of view, however, both are equally military, and each only effective if it suits the circumstances.

The third way, which is by far the most important because of the great number of cases to which it applies, is the *wearing out* of the enemy. We choose this expression not merely to give a definition, but because it represents the thing exactly and is not so figurative as it may at first appear. The idea of wearing out in a struggle implies *a gradual exhaustion of the physical powers and the will by the long continuance of action.*

Now, if we want to overcome the enemy by outlasting him in the struggle, we must content ourselves with small objects, for naturally a great object requires a greater expenditure of forces than a small one. However, the smallest object we can propose is pure resistance, a combat without any positive intention. In this case our means would be relatively at their maximum and the result most clearly assured. How far can this negative mode of proceeding be carried? Certainly not to absolute passivity, for mere endurance would cease to be a combat; resistance is an active thing, and by it so much of the enemy's force must be destroyed that he must abandon his intention. That alone is what we aim at in each single act, and therein lies the negative character of our intention.

No doubt this negative intention, as a single act, is not so effective as a positive one would be in the same direction, assuming that it succeeded. However, the difference in favor of the negative intention lies precisely in the fact that it succeeds more easily and therefore offers greater certainty of success. What it loses in effectiveness in its single act it must gain through the duration of the struggle, and thus the negative intention—which constitutes the essence of pure resistance—is also the natural means for overcoming the enemy by wearing him out.

The soldier is levied, clothed, armed, trained—he sleeps, eats, drinks, marches—merely to fight at the right place and the right time.

The difference between *offensive* and *defensive,* which governs the whole province of war, originates in this negative intention. All the advantages are deduced from it. The stronger form of combat, the defensive, thus expresses that philosophical dynamic law establishing an inverse relationship between the magnitude and the certainty of success.

If, then, the negative intention—the concentration of all means in pure resistance—affords a superiority in combat, and if this superiority is sufficient to *balance* whatever superiority in numbers the enemy may have, then the mere *duration* of the struggle will suffice gradually to bring the loss of force on the side of the enemy to the point at which his political object can no longer be an adequate equivalent. At this point he must give up the contest. We therefore see that the method of tiring out the enemy characterizes the great number of cases in which the weaker resists the stronger. . . .

. . . We see then that there are many ways to the object in war. The defeat of the enemy is not always necessary. The destruction of the enemy's military forces, the conquest of enemy provinces, the mere occupation, the mere invasion of them, enterprises aimed at political relations, and lastly a passive expectation of the enemy's onset—all these are means which, each in itself, may be used to overcome the enemy's will, according as the peculiar circumstances of the case lead us to expect more from the one than from the other. We could add to these a whole class of objects, as shorter ways of gaining our aim, which we might call arguments *ad hominem.* In what field of human affairs do sparks of personality that overcome all material circumstances fail to appear? Least of all, surely, can they fail to appear in war, where the personalities of the combatants play so important a part, both in the cabinet and in the field. We limit ourselves to pointing this out, as it would be pedantry to try to classify these means. Including them, we may say that the number of possible ways of attaining our objective rises to infinity.

To avoid underestimating the value of these various shorter ways to our object, we have only to recall the diversity of political objects that may cause a war, or to measure at a glance the distance that separates a death-struggle for political existence from a war which a forced or tottering alliance accepts as a disagreeable duty. Between these two, innumerable gradations occur in practice. If we reject one of these gradations, we might with equal right reject them all, which would be to lose sight of the real world entirely.

In general, that is the substance of the aim to be pursued in war; let us now turn to the means. There is only one means: *combat.* Regardless of the number of things that are not combat which may introduce themselves, it is always implied in the conception of war that all the effects manifested in it have their origin in combat. That this must always be so, even in the greatest diversity and complication of reality, can be proved very simply. All that occurs in war takes place through military forces; where military forces are used, the idea of combat must underlie everything. All that relates to military forces—their creation, maintenance and employment—belongs to warfare. But creation and maintenance are only the means; employment is the object.

Combat in war is not a combat of individual against individual, but an organized whole made up of many parts. In this great whole we may distinguish units of two kinds, the one determined by the subject, the other by the object. In an army the mass of combats ranges itself always into an order of new units which, again, form members of a higher order. The combat of each of these members also forms, therefore, a more or less distinct unit. Moreover, the purpose of the combat—its object—makes it a unit.

Now, to each of these distinguishable units in combat we attach the name of engagement.

If the idea of combat underlies every employment of armed forces, then the employment of armed forces in general is nothing more than the determining and arranging of a certain number of engagements.

Every military activity therefore relates, directly or indirectly, to the engagement. The soldier is levied, clothed, armed, trained—he sleeps, eats, drinks, marches—*merely to fight at the right place and the right time.*

If, therefore, all the threads of military activity terminate in the engagement, we shall be able to grasp them all when we settle the arrangement of the engagements. The effects proceed only from this arrangement and its execution, never directly from the conditions preceding them. Now, in the engagement all activity is directed to the destruction of the enemy, or rather of his *ability to fight,* for this is inherent in the concept of the engagement. The destruction of the enemy's armed forces is, therefore, always the means to attain the object of the engagement.

The object of the engagement may likewise be the mere destruction of the enemy's armed forces, but that is not by any means necessary and it may be something quite different. Whenever, for instance, the defeat of the enemy is not the only means to attain the political object, whenever there are other things which may be pursued as an aim in war, then it follows that these other things may become the object of particular acts of war, and therefore the object of engagements.

Even those engagements which, as subordinate acts, are in the strict sense devoted to the destruction of the enemy's armed forces need not have that destruction as their immediate object.

If we consider the complexity of a great armed force, of the quantity of details that come into play when it is employed, we can understand that the combat of such a force must also acquire complexity. There may and must arise for single parts of the force a number of objects which are not themselves the destruction of the enemy's armed forces, and while they certainly contribute to the increase of that destruction they do so indirectly. If a battalion is ordered to drive the enemy from a hill or a bridge, for example, the occupation of this position is usually the real object, and the destruction of the enemy forces there only a means or a secondary matter. If the enemy can be driven off by a mere demonstration, the object is nevertheless attained, but the hill or bridge will be occupied only to cause thereby a greater destruction of the enemy's armed forces. If this is the case on the field of battle, so much more must it be so in the whole theater of war, where not merely one army is opposed to the other, but one country to the other. Here the number of possible relations, and consequently of possible combinations, is greatly multiplied, the diversity of arrangements increased, and—because of the gradation of objects, each subordinate to the other—the first means further removed from the final object.

For many reasons, therefore, it is possible that the destruction of the forces immediately opposed to us is not the object of an engagement but only a means. In all such cases, it is no longer a question of making this destruction complete, for the engagement is here nothing but a *trial* of strength. It has no value in itself, but only in that of its decision.

In cases where the strengths are very unequal, a measure of them can be obtained by mere estimation. In such cases the weaker force will at once give way, and the engagement will not take place.

If the object of an engagement is not always the destruction of the enemy forces engaged in it, and if its object can often be attained as well without the engagement taking place at all—merely by the estimated result of it and the circumstances to which this estimated result gives rise—we can understand how whole campaigns can be carried on with great activity without the actual engagement playing any notable part in them.

That this may be so, history proves by a hundred examples. In how many of the cases the bloodless decision was justified—that is, did not involve a self-contradiction—and whether some of the reputations gained in them would stand criticism, we shall leave undecided, for our present concern is only to show the possibility of such a course of events in war.

We have only one means in war, the engagement; but this means, by the multiplicity of ways in which it can be employed, leads us into all the various paths of which the multiplicity of its objects permits, so that we seem to have achieved nothing. But this is not the case, for from this unity of means proceeds a thread which may be followed as it runs through the whole web of military activity and which, indeed, holds it together.

We have considered the destruction of the enemy's forces as one of the objects which may be pursued in war but have left undecided what importance should be assigned to it in comparison with the other objects. In single instances it will depend on circumstances, and as a general principle we have left its value undetermined. We are brought once more to this point, and we shall gain an insight into the value which must necessarily be accorded to it.

The engagement is the sole effective activity in war. In the engagement the destruction of the opposing force is the means to the end. This is true even if the engagement does not take place, because in that case the decision derives from the supposition that this destruction is to be regarded as beyond doubt. It follows, therefore, that the destruction of the enemy's armed forces is the foundation stone of all actions in war, the ultimate support of all combinations. All action, therefore, takes place on the assumption that if the decision by force of arms which lies at its foundation should actually take place, it would be a favorable one. The decision by arms is, for all operations in war, great and small, what cash payment is in bill transactions. However remote these relations may be, however seldom the settlements may take place, they must eventually be fulfilled.

If the decision by arms lies at the foundation of all combinations, it follows that, through a fortunate decision by arms, our opponent can make any one of these combinations impracticable. This applies not only to the decision on which our combination directly rests, but also to any other decision, provided that it is of sufficient importance. For every important decision by arms—that is, destruction of the enemy's forces—reacts upon all preceding it because, like a liquid, they tend to bring themselves to a level.

The destruction of the enemy's forces thus always appears as the superior and most effectual means, to which all others must give way.

It is, however, only when there is an assumed equality in all other conditions that we can ascribe to the destruction of the enemy's forces a higher efficacy. It would, therefore, be a great mistake to conclude that a blind dash must always gain the victory over cautious skill. An unskillful dash would lead not to the destruction of the enemy's forces but of our own, and therefore is not what is meant here. The higher efficacy belongs not to the *means* but to the *end,* and we are only comparing the effect of one realized end with the other.

If we speak of the destruction of the enemy's forces, we must expressly point out that nothing obliges us to confine this idea to the physical forces. On the contrary, the moral is necessarily implied as well, because both are interwoven with each other in the most minute details, and therefore cannot be separated. It is precisely in connection with the inevitable effect of a great act of destruction (a great victory) upon all other decisions by arms, that this moral element is most fluid, and therefore flows most easily through all the parts. The superior value of destruction of the enemy's forces over all other means must be weighed against the expense and the risks which this entails. It is only to avoid these that other methods are employed. That the means to destruction of the enemy's forces must be costly stands to reason, for, other things being equal, the wastage of our own forces is always the greater the more our aim is directed toward destruction of the enemy's forces. The risk of this means lies in the fact that the greater efficacy we seek recoils upon ourselves in the event of failure, and thus brings more disastrous consequences.

Other methods are therefore less costly when they succeed, less risky when they fail; but this necessarily implies the condition that the enemy employ the same methods. If the enemy should choose the method of the great decision by arms, our own method *must on that account be changed against our will into a similar one.* Everything depends, therefore, on the issue of the act of destruction: other things being equal, we must be at a disadvantage in all respects because our intentions and our methods had been directed partly to other things, which was not the case with the enemy. Two different objects of which one is not part of the other exclude each other, and therefore a force which is applied to attain one cannot simultaneously serve the other. If, therefore, one of the two belligerents is determined to seek the great decision by arms, he has a high probability of success as soon as he is certain that the enemy does not want it but seeks a different object. Anyone who seeks any such other object can reasonably do so only on the assumption that his enemy has as little intention as he has of seeking the great decision by arms.

But what we have here of another direction of intentions and forces relates only to other *positive objects* which we may select in lieu of the destruction of the enemy's forces, and not by any means to pure resistance, which may be adopted to exhaust the enemy's strength. In pure resistance the positive intention is lacking, and therefore our forces cannot in this case be directed to other objects, but can only be confined to defeating the intentions of the enemy.

We have now to consider the opposite of the destruction of the enemy's forces; that is, the preservation of our own. These two efforts always go together, as they react upon each other. They are integral parts of one and the same intention, and we have only to examine what effect is produced when one or the other has the predominance. The endeavor to destroy the enemy's forces has a positive object and leads to positive results, of which the final aim would be the conquest of the enemy. The preservation of our own forces has a negative object, and thus leads to defeat of the enemy's intentions: that is, to pure resistance—of which the ultimate aim can only be to prolong the contest so that the enemy exhausts himself in it.

The effort with a positive object calls into existence the act of destruction; the effort with the negative object awaits it.

How far this waiting should and may be carried we shall consider more particularly in the theory of attack and defense. We must here content ourselves with saying that the waiting must be no passive endurance, and that in the action connected with it the destruction of the enemy forces engaged in the conflict may be the aim just as well as anything else. It would therefore be a great error in fundamental principles to suppose that the negative effort must preclude our choosing the destruction of the enemy's forces as the object, and that we must prefer a bloodless decision. The advantage which the negative effort affords may certainly lead to that, but only at the risk of its not being the most suitable method, a question which depends on totally different conditions, resting not with ourselves but with our opponent. This other bloodless way thus cannot in any way be considered as the natural means of satisfying our predominating anxiety to preserve our forces. On the contrary, if such a course did not fit the circumstances we should be much more likely to bring them to utter ruin. Very many generals have fallen into this error and have been brought to ruin by it. The only necessary effect of the advantage of the negative object is delay of the decision, so that the defender takes refuge in awaiting the decisive moment. The consequence of this is usually the *putting back of the action* in time and, so far as space is connected with it, also in space, so far as circumstances permit. If the moment has arrived when this could no longer be done without overwhelming disadvantage, the advantage of the negative effort must be considered as exhausted. The effort for the destruction of the enemy's force, which was kept back by a counterpoise but never discarded, then comes forward unchanged.

The foregoing reflections show that in war there are many ways to the attainment of the political object, but that the only means is the engagement. Consequently, everything is subject to a supreme law: *the decision by arms*. When this is actually demanded by the enemy, such an appeal can never be refused. A belligerent who proposes to take another way must be sure his opponent will not make this appeal, or he will lose his case before this supreme court. Accordingly, among all the objects which may be sought in war the destruction of the enemy's forces appears always to be the one that overrules all others.

. . . When political objects are unimportant, motives weak and the passions of the forces slight, a cautious commander may try all sorts of ways by which, without great crises and bloody solutions, he may twist himself into a peace through the weaknesses of his opponent in the field and in the cabinet. We have no right to find fault with him if the assumptions on which he acts are well founded and promise success, but we must still require him to remember that he is treading a slippery path upon which the God of War may surprise him. He must always keep his eye on the enemy lest he have to defend himself with a dress rapier if the enemy takes up a sharp sword.

The consequences of the nature of war, how ends and means act in it, how in practice it deviates now more and now less from its original strict theoretical conception, fluctuating backwards and forwards yet always remaining under that strict conception as under a supreme law—all this we must keep constantly in mind in the consideration of each of the succeeding subjects, if we are to understand rightly their true relations and proper importance, and not become incessantly involved in the most glaring contradictions with reality. . . .

III

THE GENIUS FOR WAR

EVERY SPECIAL CALLING IN life, if it is to be pursued with a certain measure of perfection, demands special qualities of intellect and temperament. When these are of a high order, and manifest themselves by extraordinary achievements, the mind to which they belong is accorded the term "genius."

We know quite well that this word is used with meanings which vary greatly. In our discussion we shall adhere to the ordinary meaning of the term, and understand by "genius" a very superior mental capacity for certain activities. We are not concerned here with the kind of genius represented by a very superior talent, or genius properly so called, for that is a conception which has no defined limits. We are, instead, considering all the combined tendencies of the mind and soul toward military activity, and these we may regard as the *essence of military genius*. We say "combined," for military genius consists not of a single capacity for war, but rather of a *harmonious combination of powers,* in which one may predominate but none may be in opposition. . . .

The more the military activity predominates in a people, and the fewer other activities it pursues, the more prevalent must military genius be. However, this merely determines its prevalence, and not its degree, for the latter depends on the mental and moral development of the people. Among uncivilized peoples we never find a really great general, and very seldom what we can properly call a military genius, because that demands a development of the intellect that uncivilized peoples cannot have. On the other hand, it is obvious that civilized peoples can also have a more or less warlike tendency. The more this is the case, the more frequently will the military spirit be found in individuals in their armies. As this now coincides with the higher degree of civilization, such peoples always provide the most brilliant examples of military achievement, as the Romans and the French have shown.

Mere intellect, however, is not quite courage, for we often see the cleverest people devoid of resolution.

From this we may at once infer the importance that the intellectual powers have in superior military genius. We shall now examine this point more closely.

War is the province of danger, and therefore *courage* is, above all else, the first quality of a warrior. Courage is of two kinds; first, physical courage, or courage in the presence of danger to the person; next, moral courage, or courage in the presence of responsibility, whether before the judgment seat of an external authority or before that of the internal authority of conscience. We speak here only of the first.

Physical courage is, again, of two kinds. First, it may be indifference to danger, whether proceeding from the way the individual is constituted, from contempt of death, or from habit. In any of these cases it is to be regarded as a permanent condition. Secondly, physical courage may proceed from positive motives, such as personal pride, patriotism, or enthuiasm of any kind. In this case courage is not so much a permanent condition as an emotion.

We can readily understand that the two kinds act differently. The first kind is more certain because, having become second nature, it never deserts a man; the second often leads him farther. There is more of firmness in the first, boldness in the second. The first leaves the judgment cooler; the second raises its power at times, but also bewilders it. The two combined comprise the most perfect form of courage.

War is the province of physical exertion and suffering. In order not to be completely overcome by these, a certain strength of body and spirit is required which, whether natural or acquired, produces indifference to them. With these qualifications, under the guidance of ordinary common sense, a man is already a good instrument for war, and these are the qualifications commonly found among wild and half-civilized peoples. If we go further into the demands which war makes upon its votaries, we find intellectual qualifications predominating. War is the province of uncertainty; three-fourths of the things upon which action in war is calculated lie hidden in a fog of uncertainty. A fine and penetrating intellect is thus required to feel out the truth with instinctive judgment.

War is the province of chance. In no other sphere of human activity must such a margin be left for this intruder. It increases the uncertainty of every circumstance and deranges the course of events.

Because of these continual incursions of chance, and the uncertainty of all reports and assumptions, the person acting in war constantly finds things different from his expectations. This inevitably influences his plans, or the expectations connected with his plans. . . .

To get safely through this perpetual conflict with the unexpected, two qualities are indispensable; first, an intellect which, even in the midst of this obscurity, is not without some traces of inner light which lead to the truth; second, the courage to follow this faint light. The first is figuratively expressed by the French phrase *coup d'oeil;* the second is resolution. . . .

If we reduce the concept of *coup d'oeil* to its essence, it amounts simply to the rapid recognition of a truth which to the ordinary mind is not discernible at all or becomes so only after long examination and reflection.

Resolution is an act of courage in a single instance and, if it becomes a characteristic trait, a habit of the mind. Here we do not mean courage in facing physical danger, but courage in facing responsibility, and therefore to a degree in facing moral danger. This has often been called *courage d'esprit,* on the ground that it springs from the intellect. However, mere intellect is not quite courage, for we often see the cleverest people devoid of resolution. The intellect must first, therefore, awaken the feeling of courage to be supported by it, because in emergencies man is governed more by his emotions than his thoughts. . . .

The resolution required to overcome doubts can thus be called forth only by the intellect, and in fact by a quite special direction of it. This special direction of the intellect, which conquers the fear of wavering or hesitating and every other fear in man, is what constitutes resolution in strong minds. For that reason, men of little intellect can never be resolute, in our sense of the word. In difficult situations they may act without hesitation, but they do so then without reflection, and a man who acts without reflection is, of course, not torn asunder by doubt. The course of action selected may even, now and then, turn out to be correct; however, it is the average result which indicates the existence of military genius. Should our assertion seem strange to anyone because he knows many a resolute hussar officer who is no deep thinker, we remind him that the question here is of a special direction of the intellect and not a capacity for deep meditation. We believe, therefore, that resolution is indebted to a special direction of the intellect for its existence, a direction which belongs to a strong mind rather than a brilliant one. . . .

If we take a comprehensive view of the four components of the atmosphere of war—danger, physical effort, uncertainty, and chance—it is readily understood that a great moral and mental force is needed to cope with these baffling elements. We find historians and military chroniclers describing this force as *energy, firmness, staunchness, strength of mind and character*. . . .

Energy in action expresses the strength of the motive by which the action is called forth, whether the motive originates in an intellectual conviction or an emotional impulse. The latter, however, can hardly ever be absent where great force is to be shown. . . .

Firmness denotes the resistance of the will to the force of a single blow, *staunchness* its resistance to a series of blows. Close as the analogy between the two is, there is still a notable difference between them which cannot be mistaken, because firmness against a single powerful impression may have its root in the mere strength of a feeling, but staunchness must be supported more by the intellect. The longer an action lasts, the more deliberate it becomes, and from this deliberation staunchness partly derives its power.

If we turn now to *strength of mind or of character,* the first question is: what are we to understand thereby?

Obviously it is not violence in expressions of feeling, or proneness to strong emotion, for that would be contrary to all the usage of language, but the power of listening to reason even in the midst of the most intense excitement, in the storm of the most violent emotions. We doubt that this power depends on intellect alone. We believe we are nearer the truth if we assume that the power we call self-command, the power of submitting oneself to the control of the intellect, even in moments of the most violent excitement, has its root in the heart. It is in fact another feeling, which in men of stout heart balances the emotions without destroying them, and it is only through this balance that mastery of the intellect is secured. We may therefore say that a stout heart is one which does not lose its balance even under the most violent excitement. . . .

By the term *strength of character,* or simply *character,* is denoted tenacity of conviction, whether that conviction is based upon our own or another's judgments and whether it is based upon principles, opinions, momentary inspirations, or any other product of the intelligence. This kind of firmness cannot, of course, manifest itself if the judgments themselves are subject to frequent change. This frequent change need not be the result of an external influence. It may arise from the continuous activity of our own intelligence, but in that case it indicates an unsteadiness peculiar to that intelligence. Obviously a man who changes his views every moment cannot be said to have character, however much such changes may proceed from himself. . . .

In war, because of the many powerful impressions to which the mind is exposed, and the uncertainty of all knowledge and all judgments, more things occur to distract a man from the road he has entered upon, to make him doubtful of himself and others, than in any other human activity. The sight of suffering and danger easily leads to the emotions gaining ascendancy over intellectual conviction, and in the twilight which surrounds everything it is so difficult for a judgment to be clear and profound that to change is more understandable and more pardonable. We have always only conjectures or guesses at truth to act upon. For that reason, differences of opinion are nowhere so great as in war, and the stream of impressions acting counter to one's convictions never ceases to flow. Even the greatest intellectual impassivity is scarcely proof against them, because the impressions are too strong and vivid and are always directed at the emotions.

When the judgment is clear and profound, none but general principles and views of action governing it from a higher standpoint can be the result. On them the opinion on the particular case immediately under consideration lies, so to speak, at anchor. The difficulty is to hold fast to these results of previous reflections in opposition to the stream of opinions and phenomena which the present brings with it. Between the particular case and the principle, there is often a wide gap which cannot be bridged by a visible chain of conclusions, and where a certain belief in oneself is necessary and a certain amount of scepticism is serviceable. Here often nothing will help us but an imperative maxim which, independent of thought, controls it. The maxim is: In all doubtful cases adhere to our first opinion and do not relinquish it until a clear conviction forces us to do so. By giving preference, in doubtful cases, to our previous convictions, and by adhering to them, our actions acquire that stability and consistency which make up what we call character.

It is easy to see how greatly a well-balanced temperament promotes strength of character; that, too, is why men of great moral strength generally have a great deal of character.

Strength of character leads us to a degenerate form of it—*obstinacy.*

Obstinacy is not a fault of the intellect. We use the term as denoting resistance to our better judgment, and that cannot be located, without involving us in a contradiction, in the intellect, which is the capacity of judgment. Obstinacy is a *fault of temperament.* This inflexibility of will and impatience of contradiction find their origin only in a particular kind of egotism, which sets above every pleasure that of governing itself and others solely by its own caprice. We would call it a form of vanity if it were not, of course, something better; vanity is satisfied with the appearance but obstinacy rests upon the enjoyment of the thing.

We say, therefore, that strength of character becomes obstinacy as soon as resistance to an opposing judgment proceeds not from a better conviction or reliance upon a higher principle, but from a feeling of opposition. If this definition is of little practical assistance, it will still prevent obstinacy from being considered as merely strength of character intensified. It is something essentially different; something which, it is true, lies close to and often borders upon strength of character, but is at the same time so little an intensification of it that there are very obstinate men who, from lack of intelligence, have little strength of character. . . .

If now, in conclusion, we ask what kind of intellect is closely associated with military genius, then a glance at the subject as well as experience will tell us that searching, rather than creative, minds, comprehensive minds rather than such as pursue one special line, cool, rather than fiery, heads are those to which in time of war we should prefer to trust the welfare of our brothers and children, the honor and safety of our country.

IV

ON PHYSICAL EFFORT IN WAR

If no one were allowed to pass an opinion on the events in war except at a moment when he is benumbed by frost, or suffocating with heat and thirst, or overwhelmed by hunger and fatigue, we should certainly have fewer judgments that would be correct objectively. They would, however, at least be correct subjectively; that is, they would contain the exact relationship between the person making the judgment and the object. We recognize this when we see how modestly subdued, how spiritless and humble, is the judgment passed on the results of untoward events by those who were eyewitnesses to them, especially if they have been involved in them. This is, in our opinion, a criterion of the influence which physical effort or fatigue exerts, and of the consideration which must be given to it in forming a judgment. In war there are many things for which no objective standard of value exists; physical effort must above all be included among them. Provided that it is not wastefully employed, it is a major factor in the efficiency of all forces and no one can say precisely how far it can be carried. The remarkable thing is that, just as only a strong arm enables the archer to draw his bow, so it is only a strong spirit in the leader that can exact the utmost from the forces of his army. . . .

Just as only a strong arm enables the archer to draw his bow, so only a strong leader can exact the utmost from his army.

Although here the question is strictly of the effort required by a general from his army, by a leader from his subordinates, and thus of the courage to demand it and of the art of maintaining it, still the physical effort of the leader and of the general himself must not be overlooked. Having brought the analysis of war conscientiously up to this point, we must also take account of the importance of this small remaining element.

We have spoken of physical effort because, like danger, it belongs to the fundamental causes of friction in war, and because its indefinite quality makes its friction difficult to calculate. . . .

V

INFORMATION IN WAR

By the word "information" we mean all the knowledge we have of the enemy and his country; therefore, in fact, the foundation of all our plans and actions. Let us consider the nature of this foundation, its unreliability and uncertainty, and we shall soon feel what a dangerous edifice war is, how easily it may fall to pieces and bury us in its ruins. . . .

A great part of the information in war is contradictory, a still greater part is false, and by far the greatest part is somewhat doubtful. This requires that an officer possess a certain power of discrimination, which only knowledge of men and things and good judgment can give. The law of probability must be his guide. This is difficult even in the pre-war plans, which are made in the study and outside the actual sphere of war. It is enormously more difficult when, in the turmoil of war, one report follows hard upon another. It is fortunate if these reports, in contradicting each other, produce a sort of balance and thus demand further examination. It is much worse for the inexperienced when chance does not render him this service, but one report supports another, confirms it, magnifies it, continually paints with new colors, until urgent necessity forces from him a decision which will soon be disclosed as folly, all these reports having been lies, exaggerations, and errors.

In a few words, most reports are false, and the timidity of men gives fresh force to lies and untruths. As a general rule, everyone is more inclined to believe the bad than the good. . . . The leader, firm in reliance on his own better convictions, must stand fast like the rock on which the wave breaks. The role is not an easy one; he who is not by nature of a buoyant disposition, or has not been trained and his judgment matured by experience in war, may let it be his rule to do violence to his own inner conviction by inclining from the side of fear to the side of hope; only by that means will he be able to preserve a true balance.

This difficulty of seeing things correctly, which is one of the greatest sources of friction in war, makes things appear quite different from what had been expected. The impression of the senses is stronger than the force of ideas resulting from deliberate calculation, and this goes so far that probably no plan of any importance has ever been executed without the commander having to overcome fresh doubts during the first moments of its execution. Ordinary men, who follow the suggestions of others, generally, therefore, become undecided on the field of action; they think they have found the circumstances different from what they had expected, all the more so, indeed, since here again they yield to the suggestions of others. But even the man who has made his own plans, when he comes to see things with his own eyes, often will think he has erred. Firm reliance upon himself must make him proof against the apparent pressures of the moment. His first conviction will in the end prove true, when he extends his horizons and looks beyond the foreground scenery, with its exaggerated shapes of danger, which fate pushes on to the stage of war. This is one of the great gulfs that separate *conception* from *execution*.

VI

FRICTION IN WAR

WITHOUT PERSONAL KNOWLEDGE OF war, we cannot perceive where its difficulties lie, nor what genius and the extraordinary mental and moral qualities required in a general really have to do. Everything seems so simple, all the kinds of knowledge required seem so plain, all the combinations so insignificant, that in comparison with them the simplest problem in higher mathematics impresses us with a certain scientific dignity. If we have seen war, all becomes intelligible, yet it is extremely difficult to describe what brings about this change, to name this invisible and universally operative factor.

Everything is very simple in war, but the simplest thing is difficult. These difficulties accumulate and produce a friction beyond the imagination of those who have not seen war. . . . The influence of innumerable trifling circumstances, which cannot be properly described on paper, depresses us, and we fall short of the mark. A powerful iron will overcomes this friction; it crushes the obstacles, but at the same time the machine along with them. Like an obelisk toward which the principal streets of a town converge, the strong will of a proud spirit stands prominent and commanding in the middle of the art of war.

Friction is the only conception which, in a fairly general way, corresponds to the distinction between real war and war on paper.

The military machine, the army and all belonging to it, is fundamentally simple, and thus appears easy to manage. But let it be borne in mind that no part of it consists of one piece, that it is composed entirely of individuals, each of whom maintains his friction in all directions. Theoretically it all sounds very well: the commander of a battalion is responsible for the execution of the order given; and as the battalion by its discipline is cemented into one piece, and the chief must be a man of recognized zeal, the wheel turns on an iron bearing with little friction. In reality, however, it is not so, and all that is exaggerated and false in this conception manifests itself at once in war. The battalion is always composed of a number of men, of whom, if chance so wills, even the most insignificant is able to cause delay or some irregularity. The danger that accompanies war, the physical effort it demands, intensify the evil so greatly that they must be regarded as its most significant causes.

Everything is very simple in war, but the simplest thing is difficult.

This enormous friction is not concentrated, as in mechanics, in a few points; it is, therefore, everywhere brought into contact with chance, and thus produces incidents which are impossible to foresee, simply because it is largely to chance that they belong. . . .

Action in war is movement in a resistant medium. Just as a man immersed in water is unable to perform with ease and regularity the simplest and most natural of movements, that of walking, so in war, with ordinary powers one cannot keep even the line of mediocrity. This is why the correct theorist is like a swimming master, who teaches on dry land movements which are required in water, which must appear ludicrous to those who forget about the water. This is also why theorists who have never plunged in themselves, or who cannot deduce any generalizations from their experience, are unpractical and even absurd, because they teach only what everyone knows—how to walk.

Further, every war is rich in individual phenomena. It is therefore an unexplored sea, full of rocks which the general may suspect but which he has never seen and round which he must steer in the night. If a contrary wind also springs up—if some great chance event declares against him—then the most consummate skill, presence of mind and effort are required, while to the distant observer everything seems to be running like clockwork. The knowledge of this friction is a major part of that often boasted experience which is required in a good general. Certainly the best general is not one who gives the largest place to this knowledge and who is most overawed by it; this constitutes the class of over-anxious generals, of whom there are many among the experienced. Nevertheless, a general must be aware of it in order to overcome it, where this is possible, and to avoid expecting in his operations a degree of precision which this friction precludes. Besides, knowledge of war's friction can never be learned theoretically. Even if it could, there would still be lacking that practiced judgment which we call instinctive, and which is always more necessary in a field full of innumerable small and diversified objects than in great and decisive cases in which our judgment may be aided by consultation with others. Just as a man of the world, in whom judgment has become ingrained as a habit, speaks and acts and moves only as befits the occasion, so only the officer experienced in war will always, in matters great and small, at every pulsation of the war, decide and determine suitably to the occasion. Through this experience and practice the thought comes into his mind of itself: this is the right decision and that is not. He will, therefore, not easily place himself in a weak or exposed position, a thing which, if it occurs often in war, shakes all the foundations of confidence and becomes extremely dangerous.

It is, therefore, what we have here called friction that makes that which appears easy actually difficult. As we proceed, we shall often meet with this subject again, and it will become plain that, in addition to experience and a strong will, there are still many other rare qualities of mind required to make a distinguished general.

VII

CONCLUDING REMARKS ON BOOK I

THOSE THINGS WHICH COMBINE to make war a resistant medium for every activity we have designated as danger, physical effort, information, and friction. In their hindering effects they may be included again in the idea of a general friction. Now, is there no oil capable of diminishing this friction? Only one, and that one is not always available at the will of the commander or his army. It is the habituation of the army to war.

Habit gives strength to the body in great efforts, to the mind in great danger, to the judgment against first impressions. By its means a precious self-possession is gained through every rank, from the hussar and rifleman up to the general of division, which eases the task of the commander-in-chief. . . .

No general can give his army habituation to war, and peace-time maneuvers are but a weak substitute for it; weak, that is, in comparison with real experience in war, but not weak in relation to other armies in which peace-time maneuvers are limited to mere routine, mechanical exercises. To arrange the maneuvers in peace time so as to include some of these causes of friction, in order that the judgment, circumspection and even resolution of the separate leaders may be exercised, is of much greater value than the inexperienced believe. It is immensely important that the soldier, no matter what his rank, should not first encounter in war those things which, seen for the first time, surprise and confuse him. If he has met them only once before, he is half acquainted with them. This applies even to physical efforts; they must be practiced not so much to accustom the body to them, but the mind. In war the young soldier is very apt to regard unusual fatigues as the consequence of faults, mistakes, and embarrassment in the conduct of the whole, and thus to become depressed and despondent. This will not happen if he has been prepared for it beforehand in peace-time maneuvers. . . .

II

The Theory of War

VIII

BRANCHES OF THE ART OF WAR

WAR IN ITS LITERAL meaning is combat, for combat alone is the efficient principle in the manifold activity which is called war. But combat is a trial of strength of the moral and physical forces, by means of the latter. That the moral cannot be omitted is self-evident, for the state of the mind exerts a decisive influence on the forces employed in war.

The need for combat very soon led men to special inventions to gain the advantage in it. Combat has therefore undergone many changes, but regardless of how it is conducted, its conception remains unaltered, and combat constitutes war.

The inventions have been, first of all, weapons and equipment for the individual combatants. These have to be provided and their use learned before war begins. Their design is governed by the nature of the combat, but the fashioning of them is obviously a different thing from the combat itself; it is only preparation for combat, not the conduct of it. Since mere wrestling is also combat, it is quite clear that arms and equipment are not an essential part of combat.

Combat has determined everything pertaining to arms and equipment, and these in turn modify the combat. There is thus a reciprocal relationship between the two. Nevertheless, combat itself remains a quite special activity, the more so because it moves in a quite special element, the element of danger.

If, then, there is anywhere a necessity for drawing a line between two different activities, it is here. In order to see clearly the practical importance of this idea, we need only recall how often the greatest personal fitness in one field has turned out to be the most useless pedantry in the other.

It is also not difficult to separate, in treatment, one activity from the other, if we regard the armed and equipped forces as given means. In order to use these forces profitably we need know only their main results.

In its proper sense, therefore, the art of war is the art of making use of this given means in combat, and we cannot give it a better name than "the conduct of war." On the other hand, in a wider sense all activities which exist to support war—the whole process of creating armed forces, that is, the levying, arming, equipping and training of them—belong to the art of war.

To make a sound theory it is necessary to separate these two activities. . . .*

The conduct of war, therefore, is the arrangement and conduct of combat. If this combat were a single act, there would be no need for further subdivision. But, as we have shown in Book One, Chapter One, combat is composed of a more or less large number of single acts, each complete in itself, which we call engagements. From this two different activities spring: *individually arranging and conducting* these single engagements and *combining* them with one another to attain the object of the war. The former is called *tactics,* the latter *strategy*. . . .

According to our classification, therefore, tactics teaches *the use of the armed forces in engagements,* and strategy *the use of engagements to attain the object of the war*. . . .

Our classification concerns and covers only *the use of armed forces*. But there are in war a number of activities, subservient to it but still different from it, related to it sometimes more closely and sometimes less. All these activities relate to the *maintenance of the armed forces*. As the creation and training of these forces precedes their use,

* That is, to distinguish between the creation of armed forces and their use. Ed.

so their maintenance is inseparable from that use and a necessary condition of it. Strictly considered, however, all activities related thereto are always to be regarded as preparation for combat. . . . We therefore have the right to exclude them as well as other preparatory activities from the art of war in its restricted sense, from the conduct of war properly so called; we are obliged to do so if we wish to fulfill the first task of any theory: the separation of things that are unlike. Who would include in the conduct of war proper the whole catalog of things like subsistence and administration? These things, it is true, stand in a constant reciprocal relation to the use of the troops, but they are something essentially different. . . .

If we once more review the result of our reflections, then the activities belonging to war divide into two principal classes: such as are only *preparations for war* and such as are *the war itself.* This division must, therefore, also be made in theory.

The kinds of knowledge and skill involved in the preparations for war will be concerned with the creation, training and maintenance of armed forces. What general name shall be given them is a matter we leave open, but it is clear that among them are included artillery, fortification, so-called elementary tactics, the whole organization and administration of the armed forces. But the theory of war itself is occupied not with perfecting these means but with their use for the object of the war. . . . This theory will therefore deal with the engagement as the real combat, and with marches, camps and quarters as matters more or less identical with it. The subsistence of the troops will only come into consideration like *other given circumstances* in respect of its results, not as an activity belonging to the theory itself.

This art of war, in its more limited sense, is in turn divided into tactics and strategy. The former is concerned with the form of the individual engagement, the latter with its use. Both are concerned with the circumstances of marches, camps and quarters only through the engagement, and these things will be tactical or strategic, according as they relate to the form or to the significance of the engagement. . . .

IX

ON THE THEORY OF WAR

FORMERLY THE TERM "ART of war" or "science of war" was understood to mean only those branches of knowledge or skill which are concerned with material things. The design and preparation and use of weapons, the construction of fortifications, the organization of the army and the mechanism of its movements were the subjects of these branches of knowledge and skill, and the object of them all was the design of an armed force suitable for war. . . . The relation of all this to combat is very similar to the relation of the sword cutler to the fencer. The question of the employment. . . (of armed forces in war) had not been considered.

Something of the conduct of war itself, something of the action of the intellectual faculties upon the material forces under their control first appeared in the art of sieges. For the most part, however, these intellectual faculties concentrated upon new material objects, such as approaches, counterapproaches, trenches, batteries, and so forth. Every intellectual step showed itself in such a material result. . . . This kind of approach was then more or less adequate, since in this kind of war the intellect could hardly manifest itself except in material things.

Later on tactics attempted to develop a universally valid arrangement, founded upon the peculiar properties of the instrument available to it. Although this approach certainly leads to the battlefield, it excludes a free activity of the mind; on the contrary, it leads to reduction of the army to an automaton. Employing rigid formation and order of battle, and put into motion by a mere word of command, its activity was intended to proceed like clockwork.

The conduct of war properly so called—that is, the intelligent application of previously prepared means to peculiar needs—could not, it was thought, be a subject for theory, but must be left to natural talents alone. By degrees, as war passed from the hand-to-hand encounters of the Middle Ages into a more regular and composite form, stray reflections on this matter did enter into the minds of men, but for the most part they appear only in memoirs and narratives, incidentally and, in a sense, incognito.

As reflections on military events became increasingly numerous and history assumed a more critical character, an urgent need arose for a body of principles and rules which could terminate the conflict of opinions and controversy which had naturally arisen. . . .

Accordingly, an endeavor arose to establish principles, rules and even systems for the conduct of war. The attainment of a positive objective was thus proposed, without sufficient consideration of the innumerable difficulties which war presents in this connection. The conduct of war has, as we have shown, no definite limits in any direction, whereas every stem, every theoretical construction, has the limiting nature of a synthesis. The result is an irreconcilable opposition between such a theory and practice.

Writers on theory soon felt the difficulty of the subject, and considered themselves justified in again avoiding it and directing their principles and systems to material things and to one-sided activity. They wished, as in the sciences dealing with the preparations for war, to arrive at perfectly certain and positive results, and therefore only to consider what could be made a matter of calculation.

Superiority in numbers, being a material thing, was chosen from all other factors required to produce victory, because by combinations of time and space it could be subjected to mathematical laws. It was thought possible to consider it apart from all other circumstances, by assuming those to be equal on both sides and, consequently, to neutralize each other. This would have been acceptable had it been done only temporarily to gain knowledge of this one factor, according to its relations, but to do so permanently—to consider superiority in numbers as the sole law, and to see the whole secret of war in the formula: *to bring up superior numbers to a certain point at a certain time*—was a restriction absolutely untenable against the force of reality.

One theoretical treatment attempted to make the subsistence of the troops. . . the supreme arbiter in the higher conduct of war. Definite figures were arrived at, but these rested upon a number of arbitrary assumptions and could not withstand the test of practical application.

An ingenious author tried to concentrate in a single conception—that of a *base*—a whole host of things, including even some relations with mental and moral forces. The list comprised the subsistence of the army, the maintenance of the level of its numbers and equipment, the security of communications with its home country, and finally the security of retreat in case it became necessary. . .all this merely to arrive at a purely geometrical result, which is quite worthless. . . . The conception of a base is for strategy a real need, and to have conceived it is meritorious, but to make such a use of it as we have indicated is quite inadmissible and could only lead. . . in an absolutely absurd direction—to a belief, namely, in the superiority of the enveloping form of attack.

As a reaction against this false tendency, another geometric principle, that of interior lines, was then elevated to the throne. Although this principle rests on the sound foundation that the engagement is the only effectual means in war, its purely geometric nature means it is nothing but another instance of a one-sided theory which cannot govern real events.

All these attempts at theory, so far as they are analytical, can be regarded as advances in the domain of truth, but in their synthetical aspects, in their precepts and rules, they are quite useless.

They strive after determinate quantities, whereas in war all is undetermined, and the calculation must be made with variables.

They direct the attention only upon material forces, whereas military action is permeated throughout with immaterial forces and effects.

They consider the action to be on one side only, whereas war is a constant reciprocal action between the opposing forces.

Combat gives birth to the element of danger, in which all the activities of war must live and move, like the bird in the air or the fish in the water.

All that was unattainable by this treatment, which neglected all elements but one, lay outside the precincts of science. It was, according to them, the field of genius, *which raises itself above all rules*. . . . What genius does must be the best of all rules, and theory cannot do better than to show how and why it is so.

Pity the theory which sets itself in opposition to mental and moral forces! It cannot compensate for this contradiction by any humility, and the humbler it is, the sooner will ridicule and contempt drive it out of real life.

Every theory becomes infinitely more difficult the moment it touches upon the province of mental and moral quantities. Architecture and painting know exactly where they stand as long as they have only to deal with matter; there is no dispute about mechanical and optical construction. But as soon as mental and moral effects begin to operate, as soon as mental and moral impressions and feelings are to be produced, the whole set of rules dissolves into vague ideas. . . .

Now, the activity in war is never directed solely against matter; it is always simultaneously directed against the mental and moral force which gives life to matter, and to separate one from the other is impossible. But the mental and moral forces are discernible only to the inner eye, and this is different in each person, and often different in the same person at different times.

As danger is the general element in which everything moves in war, it is chiefly courage, the feeling of our own strength, that influences our judgment in different ways. It is, so to speak, the lens through which all images pass before reaching the intelligence.

Everyone knows the moral effects of a surprise, of an attack in flank or rear. Everyone thinks less of the enemy's courage as soon as he turns his back, and everyone ventures much more in pursuit than when pursued. Everyone judges his opponent by his reputed talents, by his age and experience, and acts accordingly. Everyone looks critically at the spirit and morale of his own and the enemy's troops. All these and similar effects in man's mental and moral nature have been proved by experience and are constantly recurring. They therefore merit our consideration as real quantities. What could we do with a theory that failed to consider them?

. . . In order to comprehend clearly the difficulty of the problem involved in a theory of the conduct of war, and thence to deduce the necessary characteristics of such a theory, we must examine more closely the chief characteristics which comprise the nature of military action.

The first of these characteristics are moral forces and effects.

Combat is, in its origin, the expression of *hostile feeling,* but in the great combats which we call wars, the hostile feeling often becomes merely a hostile *intention,* and there is usually no hostile feeling of individual against individual. . . . National hatred, which is seldom lacking in our wars, becomes a more or less powerful substitute for personal hostility of individuals. Where this is also absent, and initially no feeling of hostility exists, a hostile feeling is kindled by the combat itself; an act of violence committed upon us. . . will excite in us the desire to retaliate and be avenged. . . . This is human—animal, if you prefer—but it is a fact. Theory is very apt to look upon combat as a trial of strength, an isolated phenomenon in which the feelings have no part. This is one of the thousand errors which theorists deliberately commit, because they fail to see the consequences of them.

In addition to the excitation of feelings aroused by the nature of combat itself, there are others which do not essentially belong to it but which, because of their relationship, easily unite with it—ambition, love of power, enthusiasms of every kind, and so forth.

The second characteristic is the consciousness of danger. Combat gives birth to the element of danger, in which all the activities of war must live and move, like the bird in the air or the fish in the water. The effects of danger, however, all pass on to the emotions, either directly—that is, instinctively—or through the intelligence. The effect, in the first case, would be a desire to escape the danger and, if that cannot be done, a feeling of fear and anxiety. If this effect does not occur, then it is *courage* which acts as the counterpoise to instinct. Courage, however, is by no means an act of the intelligence, but rather a feeling, like fear. Fear is directed to physical preservation, courage to moral preservation. Courage is a nobler instinct. Because it is so, it cannot be used like a lifeless instrument, which produces its effects in a degree precisely predetermined. Courage is, therefore, no mere counterpoise to neutralize the effects of danger, but a special quantity in itself.

In order to estimate correctly the influence of danger upon leaders in war, we must not limit its sphere to the physical dangers of the moment. It dominates the leader by threatening not only him personally, but also all those entrusted to him, not only at the moment in which it is actually present, but also—through the imagination—at all other moments. Lastly, it dominates not only directly and through itself, but also indirectly by the responsibility which makes it bear with a tenfold weight upon the leader's mind. Who could advise, or resolve upon a great battle, without feeling his mind more or less wrought up and paralyzed by the danger and responsibility which such a great act of decision carries in itself?

If we look upon these emotional forces which are excited by hostility and danger as belonging peculiarly to war, we do not for that reason exclude from war all the other emotions which accompany man on his journey through life. Here, too, these will often enough find room. We may say, it is true, that in this serious duty of life (war) many a petty play of emotion is silenced, but that holds good only for those in the lower ranks who, hurried from one state of exertion and danger to another, lose sight of all the other things in life, *become unused to deceit* because it is of no avail with death, and so attain that soldierly simplicity of character which has always been the best and most characteristic quality of the military profession. In the higher ranks it is otherwise, for the higher a man's rank the more he must look about him.

There interests arise on every side, and a manifold activity of the passions of good and evil. Envy and nobility of mind, pride and humility, harshness and tenderness, all may appear as active forces in this great drama.

The mental qualities of a leader, next to his moral qualities, are likewise of great importance. From an imaginative, volatile, inexperienced mind other things can be expected than from a calm and powerful intellect.

This great diversity in mental and moral individuality, the influence of which must be considered as principally felt in the higher ranks, because it increases as we go upward, is what produces the diversity in ways, noticed in the first book, to achieve our aim. It is also this which gives such an unequal share to the play of probability and luck in determining the course of events.

The second quality in war is quick reaction and the reciprocal action that springs from it. We do not here speak of the difficulty of estimating that reaction, for that is included in the previously mentioned difficulty of dealing with mental and moral qualities as quantities. We are concerned, rather, with the fact that reciprocal action revolts against regularity and plan. . . . It is therefore natural that in . . . war, which in its plan—built upon general circumstances—is so often thwarted by unexpected and singular events, more must generally be left to talent, and less use can be made of a theoretical *guide* than in any other business.

Lastly, the great uncertainty of all data in war is a characteristic difficulty, because all action must be directed, to a certain extent, in a mere twilight, which in addition not infrequently—like the effect of fog or moonlight—gives to things an exaggerated size and grotesque form.

What this feeble light leaves indistinct to the vision, talent must discover, or it must be left to chance. It is therefore again talent, or the favor of fortune, on which we must depend, for lack of objective knowledge.

This being the nature of the subject, we must admit that it is a sheer impossibility to erect, so to speak, a scaffolding of positive rules which would give the leader support on all sides. In all those situations in which he is thrown upon his talents, he would find himself outside of this scaffolding of rules and in opposition to it. However many-sided its construction might be, the same result, of which we have already spoken, would ensue: talent and genius would act beyond the law, and theory would become an opposite to reality. A positive system of rules is therefore impossible.

Two ways out of this difficulty present themselves.

In the first place, what we have said of the nature of military action in general does not apply in the same manner to all ranks. In the lower ranks the spirit of self-sacrifice is more required, but the difficulties which the understanding and judgment meet are infinitely less. The field of events is more confined. Ends and means are fewer. The data are more distinct, being for the most part contained in things actually visible. However, the higher we ascend, the more the difficulties increase, until in the commander-in-chief they reach their climax, so that with him almost everything must be left to genius.

Further, a division of the subject, *according to the intrinsic nature of its elements,* shows that the difficulties are not everywhere the same, but diminish the more results manifest themselves in the material world, and increase the more they pass into the mental and moral, and become motives which influence the will. It is therefore easier, by theoretical rules, to determine the order, plan and conduct of an engagement than the use to be made of the engagement. In the engagement physical weapons clash, and although mental and moral elements cannot be absent, matter must be allowed its rights. But in the effects of the engagement, when the material results become motives, we are concerned solely with mental and moral elements.

In a word: it is much easier to develop a theory for *tactics* than for *strategy*.

The second opening for the possibility of a theory lies in the viewpoint that it need not be a body of positive rules, a *guide* for action. As a general rule, whenever an activity is for the most part concerned continually with the same things, with the same ends and means, although with small variations and a corresponding variety of combinations, these things must be capable of becoming a subject of study for the reasoning faculties, by observation. Such observation is the most essential part of every *theory,* and quite properly lays title to that name. It is an analytical investigation of the subject, leading to an exact knowledge of it, and if brought to bear on experience, which in our case would be military history, to a thorough familiarity with it. . . . If an expert spends half his life in the endeavor to clarify an obscure subject in all its details, he will probably know more about it than a person who seeks mastery of it in a short time. Theory, therefore, exists in order that each person need not have to clear and plow the same ground, but may find it cleared and put in order. It should educate the mind of the future leader in war, or rather guide him in his self-instruction, but not accompany him to the field of battle, just as a sensible tutor guides and assists a youth's intellectual development without keeping him in leading-strings all his life.

If principles and rules develop from the observation that theory institutes, if the truth crystallizes into these forms, then theory will not oppose this natural law of the mind. It will rather, if the arch ends in such a keystone, bring it out more prominently, but it does so only to satisfy the philosophical law of thought. . . . For even these principles and rules serve more to determine in the reflective mind the general outlines of its accustomed movements than as signposts pointing the way to take in execution.

This point of view makes possible a satisfactory theory of the conduct of war; that is, a theory which will be useful and never opposed to reality. It will depend on intelligent handling to reconcile it so completely with practice that there will no longer be absurd differences between them.

Theory has, therefore, to consider the nature of means and ends.

In tactics the means are the armed forces which are to conduct the combat. The end is victory, which will be better defined later when we consider the engagement. Here we content ourselves by specifying the enemy's withdrawal from the field as the sign of victory. By means of this victory, strategy gains the object which it sought by means of engagement and which constitutes the real significance of it. This significance has indeed a certain influence on the nature of the victory; a victory intended to weaken the enemy's forces is a different thing from one designed merely to achieve possession of a position. The significance of an engagement can thus have a notable influence on the planning of it; consequently it will also be a subject of consideration in tactics.

. . .For strategy the tactical victory (in the engagement) is primarily only a means, and the things which should lead directly to peace are its ultimate object. The employment of the means to attain this object is influenced, as in tactics, by the accompanying circumstances, which in strategy are country and ground, the time of year, and the state of the weather.

. . .By combining these circumstances with the result of an engagement, strategy gives this result—and therefore the engagement—a special significance, assigns to it a special object. When this object is not one which leads directly to peace and is thus a subordinate object, it is also to be regarded only as a means. In strategy, therefore, we may regard successful engagements, or victories, with all their different significances, as means. When engagements are combined toward a common objective, this is also to be regarded as a means. . . . There remains, therefore, as objectives, only those things which may be supposed to lead directly to peace. . . .

The first question is: How does strategy arrive at a complete enumeration of these things? If a philosophical investigation were aimed toward an absolute result, it would become entangled in all the difficulties which exclude logical necessity from the conduct of war and the theory of it. Accordingly, it turns to experience and directs its attention to precedents provided by military history. In this way only a limited theory can be derived, appropriate only to such circumstances as military history can provide. This limitation is, in any case, unavoidable because theory must be either deduced from or compared with military history. Moreover, such a limitation is in any case more theoretical than real. One great advantage of this method is that theory cannot lose itself in hair-splitting and chimerical subtleties but must remain practical.

Another question is: How far should theory go in analysis of the means? Obviously, only so far as the different components present themselves for consideration in practice. The range and effect of different weapons are very important to tactics: their construction, although responsible for these effects, is a matter of complete indifference. The conduct of war is not the production of cannon and powder out of a given quantity of charcoal, sulphur, and saltpetre, of copper and tin; the given quantities for the conduct of war are arms in their finished state and their effects. Strategy makes use of maps without troubling itself about triangulations; it does not inquire what the institutions of a country should be and how the people should be educated and governed for best results in war. It takes these things as it finds them in the community of European states, and points out where very different conditions have a notable influence on war.

It is easy to see that, in this manner, the number of subjects for theory is greatly reduced and the knowledge required for the conduct of war greatly limited. The many kinds of expert knowledge and skill which support military activity in general, and which are necessary to put a fully equipped army in the field, join in a few major groups before they reach the point of attaining, in war, the final goal of their activity, just as the streams of a country unite into rivers before they reach the sea. Only those activities which empty directly into the sea of war need be familiar to the leader who is to direct their course.

. . .Only thus is explained how so often men have achieved great success in war and, indeed, in the higher ranks—even in supreme command—whose former pursuits had been of a totally different nature; how, indeed, the most distinguished generals have never risen from the very learned or really erudite class of officers, but have been mostly men who, from the circumstances of their position, could not have attained any great amount of knowledge. For that reason, those who have considered it necessary, or even merely useful, to begin the education of a future general by instruction in all details have always been ridiculed as absurd pedants. It is easy to show that such a course will do him harm, because the human mind is formed by the kinds of knowledge imparted to it and direction given to its ideas. Only what is great can make it great; the small can only make it small, unless the mind rejects it as something repugnant to it.

. . .In the field of military activity, the knowledge required differs according to the position of the leader. It must be directed to less important and more limited objects if he holds an inferior position, upon greater and more comprehensive things if his position is higher. There are commanders-in-chief who would not have shone at the head of a cavalry regiment, and vice versa.

Although knowledge in war is very simple—that is, directed to so few subjects, and embracing these only in their final results—to put it into practice is not very easy. We have already spoken in Book One of the difficulties to which action in war is subject. We here omit those things which can be overcome only by courage, and we maintain that the proper activity of the mind is simple and easy only in the lower ranks. It increases in difficulty as rank increases, and the highest position—that of commander-in-chief—is to be reckoned among the most difficult things of which the mind is capable.

The commander of an army need not be either a learned student of history or a publicist, but he must be knowledgeable of the higher affairs of state; he must know and be able to judge correctly regarding traditional tendencies, the interests at stake, the questions at issue, and the leading personalities. He need not be a close observer of men, a delicate dissector of human character, but he must know the character, the habits of thought, and the characteristic strengths and weaknesses of those whom he commands. . . .

The necessary knowledge for a high military position is therefore distinguished by the fact that by observation, by study and reflection, it can be gained only by a special talent, an intellectual instinct which extracts from the phenomena of life only their essence, as bees do the honey from the flowers. This instinct can be gained by experience of life as well as by study and reflection. Life with its rich teachings will never produce a Newton or a Euler, but it may well produce the higher powers of calculation for war possessed by a Condé or a Frederick. . . .

Now we have yet to consider one condition which is more necessary for the knowledge of the conduct of war than for any other, which is that it must become almost wholly part of oneself and almost completely cease to be something objective. In almost all other arts and occupations the person acting can make use of truths which he has learned only once, but in the spirit and sense of which he no longer lives, and which he extracts from dusty books. Even truths which he has in hand and uses daily may continue to be something quite external to himself. . . . But it is never so in war. The mental and moral reaction, the ever-changing form of things, makes it necessary for the person acting to carry within himself the whole mental apparatus of his knowledge and to be able, anywhere and at any moment, to produce from within himself the decision required. Knowledge must thus, by being completely assimilated into his own mind and life, be converted into real skill. That is why, with men distinguished in war, everything seems so easy and is ascribed to natural talent, as distinguished from that which is developed by observation and study.

We think that by these reflections we have explained the problem of a theory of the conduct of war, and pointed out the way to its solution.

Of the two fields into which we have divided the conduct of war, tactics and strategy, the theory of the latter contains unquestionably, as previously observed, the greatest difficulties. The first is almost entirely limited to a circumscribed field of things, but the latter, with regard to the objects leading directly to peace, opens into an undefined region of possibilities. Since it is, for the most part, only the commander-in-chief who has to keep these objects in view, the part of strategy in which he moves is also particularly subject to this difficulty.

Theory, therefore, will stop much sooner in strategy—especially where it comprehends the highest levels of decision—at the mere consideration of things, and content itself with helping the soldier to that insight into things which, blended with his whole thought, makes his course easier and surer, and never forces him into opposition with himself in order to obey an objective truth.

X

ART OF WAR OR SCIENCE OF WAR?

WHERE CREATION AND PRODUCTION are the object, there is the domain of art; where investigation and knowledge are the object, there science reigns. It is therefore obvious that it is more fitting to speak of "art of war" than "science of war."

So much for this, for we cannot do without these conceptions. But now we assert that war is neither an art nor a science in the proper sense, and it is just the setting out from that departure point in ideas which has led to a wrong direction being taken, and which has caused war to be put on a par with other arts and sciences, and led to many erroneous analogies. . . . War is a form of human intercourse. It belongs not to the province of the arts and sciences, but to that of social existence. It is a conflict of great interests which is settled by bloodshed, and only in that is it different from other conflicts. It would be better, instead of comparing it with any art, to liken it to business competition, which is also a conflict of human interests and activities; and it is still more like politics, which again, on its part, may be regarded as a kind of business competition on a great scale. Moreover, politics is the womb in which war is developed, in which its outlines lie hidden in a rudimentary state, like the qualities of living creatures in their embryos.

The essential difference lies in this: War is an activity of the will, not—like the mechanical arts—exerted upon dead matter. . . but upon a living and reacting force. How little the categories of the arts and sciences are applicable to such an activity strikes us at once, and we can at the same time understand how the constant seeking and striving after laws like those which can be evolved from the dead world of matter could not but lead to constant errors. And yet it is just the mechanical arts that some people have wanted to take as a model for constructing an art of war. Use of the fine arts as a model was quite out of the question, because these themselves still lack laws and rules. . . .

Whether such a conflict of living elements as arises and is settled in war is subject to general laws, and whether these can serve as a guide to action, will be partly investigated in this book. This much is self-evident: This subject, like any other which does not exceed our powers of understanding, may be illuminated and made more or less clear in its inner relations by an inquiring mind, and that alone is sufficient to realize the idea of a theory.

III

Of Strategy in General

XI

STRATEGY

STRATEGY HAS BEEN DEFINED in Book II, Chapter One. It is the use of the engagement to attain the object of the war. . . . It must therefore give an aim to the whole military action. This aim must be in accord with the object of the war. In other words, strategy develops the plan of the war, and to the aforesaid aim links the series of acts which are to lead to it; that is, it plans the separate campaigns and arranges the engagements to be fought in each of them. Since these are matters which, to a great extent, can only be based on assumptions, and some of these turn out to be incorrect, while a number of other decisions pertaining to details cannot be made beforehand at all, it is evident that strategy must accompany the army to the field in order to arrange particulars on the spot, and to make the modifications in the general plan which constantly become necessary. Strategy can therefore never take its hand from its work for a moment.

That this has not always been the view taken is evident from the former custom of keeping strategy with the cabinet and not with the army. Such a thing is permissible only if the cabinet remains so close to the army that it can be regarded as its chief headquarters.

. . .It is only in the highest branches of strategy that intellectual complications and a great diversity of quantities and relations are to be looked for. At this point strategy borders on politics and statesmanship, or rather it merges with them, and as we have observed, these have more influence on how much or how little is to be done than on the manner in which it is to be done. Where the latter is the principal question, as in single acts of war both great and small, the mental and moral qualities are already reduced to a very small number.

Thus, then, in strategy everything is very simple, but not on that account very easy. Once it is determined from the conditions within the state what war shall and can do, then the way to it is easy to find; but to follow that way straight forward, to carry out the plan without being obliged to deviate from it a thousand times by a thousand varying influences requires, besides great strength of character, great clearness and steadiness of mind. Out of a thousand men who are remarkable, some for mental capacity, others for penetration, others again for boldness or strength of will, perhaps not one will combine in himself all those qualities which are required to raise a man above mediocrity in the career of a general.

It may sound strange, but for all who know war in this respect, it is a fact beyond doubt, that much more strength of will is required to make an important decision in strategy than in tactics. In the latter we are carried away by the moment; a commander feels himself borne along by a powerful current, against which he dare not contend without the most destructive consequences. He suppresses his rising fears, and boldly ventures further. In strategy, where all moves much more slowly, there is much more room for our own apprehensions and those of others, for objections and remonstrances, consequently also for untimely regrets. Moreover, in strategy we do not see things with our own eyes as we do at least half of them in tactics, but everything must be conjectured and assumed. The convictions produced are therefore less powerful. The result is that most generals, when they should act, are stuck fast in bewildering doubts. . . .

In strategy, possible engagements have consequences, and must therefore be regarded as real ones. If a detachment is sent to cut off the retreat of a fleeing enemy, and the enemy surrenders without further resistance, it is to the engagement which was offered him that his surrender was due. If part of our army occupies an undefended enemy province. . . it is only through the engagement which the enemy must expect if he undertakes to recover it that we remain in possession of it.

In both cases, therefore, the possibility of an engagement has produced consequences, and has thus entered into the category of real things. . . . In this manner we see that the destruction of the enemy's military forces. . . are accomplished only through the effects of the engagement, whether it actually occurs or is merely offered and rejected.

The effects of the engagement are twofold: direct, and indirect. They are indirect if other things—things which cannot be regarded, in themselves, as destruction of the enemy's armed forces, but which are only supposed to lead to that destruction—intervene and become the object of the engagement. The possession of provinces, cities, fortresses, roads, bridges, etc., may be the immediate object of an engagement, but never the ultimate one. Things of this nature must be regarded only as means of gaining greater superiority, so that the engagement may be finally offered the opponent under conditions that make it impossible for him to accept. All these things are therefore only to be regarded as intermediate steps, leading to the effective principle but not to be confused with it. . . .

If we do not accustom ourselves to look upon war, and upon a single campaign in war as a chain composed of nothing but engagements, of which one is always the cause of the next; if we adopt the idea that the capture of certain geographical points, the occupation of undefended provinces is something in itself, we are likely to regard it as an advantage which can be picked up in passing. If we look at it so, and not as a link in a chain of events, we do not question whether this possession may not later lead to greater disadvantages. How often we find this mistake recurring in the history of war! We might say that, just as in commerce the merchant cannot set apart and place in security the gains from a single transaction, so in war a single advantage cannot be separated from the result of the whole. Just as the merchant must always operate with the whole sum of his means, so in war only the final total will decide whether any particular item is profit or loss.

But, if the mind's eye is always directed upon the series of engagements, so far as that series can be foreseen, then it is always looking in the right direction, and thereby the. . . movement of our strength acquires the energy which the occasion demands, and which must not be diverted by extraneous influences.

XII

MORAL ELEMENTS

We must return to this subject, which is touched upon in Book I, Chapter 3, because the moral elements are among the most important in war. They constitute the spirit which permeates the whole being of war. They attach themselves sooner or later, and with greatest affinity, to the will which sets in motion and guides the whole mass of forces, uniting with it to constitute one whole, because it is itself a moral element. Unfortunately, the moral elements defy book-analysis, for they can be neither counted nor classified, and must be seen and felt.

The spirit and other moral qualities of an army, a general or a government, public opinion in the provinces where the war is proceeding, the moral effect of a victory or defeat: all these are things which vary greatly in their nature. Depending on how they stand in relation to our object and our circumstances, they may also have a very different kind of influence.

Although little can be said of these things in books, still they belong as much to the theory of the art of war as everything else which constitutes war. For once more I must repeat that it is a miserable philosophy if, as was formerly done, we establish rules and principles which ignore all moral qualities, and then, as soon as these qualities appear, we begin to count the exceptions and after a fashion make them into rules; or if we resort to an appeal to genius, which is above all rules, thus implying not only that rules were made for fools, but that they must themselves be really folly.

Even if the theory of war could actually do no more than call these things to remembrance, showing the necessity of according full value to the moral qualities and always taking them into consideration, it would by so doing have extended its borders to include the whole area of immaterial forces. By establishing that point of view, it would have condemned beforehand every one who would seek to justify himself before its judgment seat by the mere physical condition of forces.

. . .Even the most uninspired theories have had to stray over into the moral kingdom; for example, the effects of a victory can never be fully explained without consideration of the moral impressions. Therefore, most of the subjects which we shall consider in this book are composed half of physical, half of moral, causes and effects. We might say that the physical are hardly more than the wooden handle, while the moral are the noble metal, the real, brightly polished weapon.

The value of the moral elements, and their frequently incredible influence, are best exemplified by history, and this is the most noble and genuine nourishment which the mind of a general can extract from it.

. . .We might go through the most important moral phenomena in war, and with all the care of a diligent professor try to impart what we could about each, good or bad. . . . We prefer to remain here more than usually incomplete and rhapsodical, content to have drawn attention to the importance of the subject in a general way, and to have indicated the spirit in which the views advanced in this book have been formed.

XIII

THE CHIEF MORAL POWERS

THE CHIEF MORAL POWERS are: *the talents of the commander, the military virtue of the army, its national feeling.* Which of these is most valuable no one can determine in a general way. It is very difficult to say anything at all concerning their strength, and still more difficult to compare the strength of one with another. The best plan is not to undervalue any of them. . . .

It is true, however, that in modern times the armies of European states have arrived very much at a par in discipline and training. The conduct of war has—as philosophers would say—developed so naturally, and become a kind of method common to all armies, that even on the part of commanders we can no longer reckon on the application of special devices in the more limited sense (such as Frederick the Second's oblique order). Hence it cannot be denied that, as matters now stand, greater scope is afforded for the influence of national spirit and of the habituation of an army to war. A long peace may again alter all this. . . .

XIV

MILITARY VIRTUE OF AN ARMY

MILITARY VIRTUE IS DISTINGUISHED from mere bravery, and still more from enthusiasm for the cause for which the war is fought. Bravery is certainly a necessary constituent of it, but just as bravery—which is ordinarily a gift—can arise in a soldier from habit and training, so with him it must also have a different direction from that which it has with other men. It must lose that impulse toward unbridled activity and manifestation of force which it has with the individual, and submit itself to demands of a higher nature, such as obedience, order, rule, and method. Enthusiasm for the cause gives life and greater fire to the military virtue of an army, but does not constitute a necessary part of it.

War is a special profession. . . . Even if all the male population of a country capable of bearing arms were to practice it, war would still be different and separate from the other activities which occupy man. To be imbued with the spirit and essence of this profession, to assimilate into our system the powers which should be active in it, to apply our intelligence to every detail of it, to gain confidence and expertness in it through exercise, to go into it heart and soul, to pass from the man into the role which is to be assigned to us in it—that is the military virtue of an army in the individual.

No matter how carefully we seek to combine the citizen and the soldier in the same individual, no matter how much we may regard wars as affairs of the whole nation, no matter how far our ideas may depart from the mercenary armies of former days, it will never be possible to dispense with the special nature of the professional routine. If that cannot be done, those who belong to the profession will always look upon themselves as a kind of guild, in the regulations, laws, and customs of which the spirit of war is predominantly expressed. And so it is in fact. Even with the most decided inclination to regard war from the highest point of view, it would be very wrong to deprecate this corporate spirit—this esprit de corps—which can and must exist more or less in every army. This corporate spirit forms the bond which unites the natural forces which are active in what we have called military virtue.

An army which maintains its formations under the heaviest fire, which is never shaken by imaginary fears, which resists with all its might any that are well-founded, which, proud in the feeling of its victories, never loses its sense of obedience, its respect for and confidence in its leaders, even in the depression of defeat; an army with its physical powers strengthened by privation and exertion; an army which regards its toils as the means to victory, not as a curse, and which is always reminded of its duties and virtues by one single idea: the honor of its arms—such an army is imbued with the true military spirit.

Only in effort does the soldier come to know his powers.

. . .Great things may be effected, soldiers may fight bravely, commanders may successfully lead standing armies, without the assistance of this military virtue; therefore, we must not say that a successful war cannot be imagined without it. We draw special attention to this point. . . so that it may not be thought that military virtue is the one and only thing. It is not. Military virtue appears as a definite moral power, the influence of which may be estimated, the strength of which may be calculated.

Having thus characterized it, we shall see what can be said about its influence, and about the means of gaining this influence.

Military virtue is everywhere for the parts what the genius of the commander is for the whole. The general can only direct the whole, not each part, and where he cannot direct the part, there military spirit must be its leader. A general is chosen by virtue of his reputation for excellent qualifications, but careful examination of qualifications decreases as we descend the scale of rank, and at the lower levels military virtue must compensate for deficiencies in individual talent. It is precisely this part that is played by the natural qualities of a people mobilized for war: *bravery, adaptability, powers of endurance and enthusiasm*. These properties may be substituted for military virtue, and vice versa, from which it may be deduced that:

1. Military virtue is a quality of standing armies only, and they require it the most. In national uprisings and in war natural qualities, which develop more rapidly in these, are substituted for it.
2. Standing armies opposed to standing armies can more easily dispense with it than can a standing army opposed to a national insurrection, for in that case the troops are more scattered and the units left more to themselves. But where an army can be kept concentrated, the genius of the general plays a greater role and makes up for what is lacking in the spirit of the army. Generally, therefore, military virtue becomes all the more necessary the more the theater of operations and other circumstances complicate the war and cause the forces to be scattered.

From these truths the only lesson to be derived is that, if an army lacks military virtue, every endeavor should be made to simplify the operations as much as possible or to double the attention paid to other parts of the military system, and not to expect from the mere name of a standing army that which only true military virtue can give.

The military virtue of an army is therefore one of the most important moral powers in war. . . .

This spirit can arise from only two sources, and these can produce it only by working together. The first is a series of campaigns and successful results; the other is the practice of often working an army to the last ounce of its strength. Only in this effort does the soldier come to know his powers. The more a general is in the habit of demanding from his troops, the more certain he is that his demands will be satisfied. The soldier is as proud of hardship overcome as he is of danger surmounted. It is, therefore, not only in the soil of incessant activity and exertion that this germ will thrive, but also in the sunshine of victory. Once it has developed into a strong tree, it will resist the fiercest storms of misfortune and defeat, and even for a time the sluggish inactivity of peace. It can therefore only be created in war, and under great generals, but no doubt it may last for several generations, even under generals of moderate capacity, and also through considerable periods of peace.

Between this extended and ennobled esprit de corps in a handful of scar-covered, war-hardened veterans and the self-esteem and vanity of a standing army which is held together merely by the bond of service regulations and a drill book, there is no comparison. A certain grim severity and strict discipline may prolong the life of military virtue, but cannot create it. These things have a certain value, but they must not be over-rated. Order, smartness, good will, also a certain degree of pride and high morale are qualities of an army trained in peace time which are to be valued, but which cannot stand alone. The whole maintains the whole, and as with a glass too quickly cooled, a single crack breaks the whole mass. Above all, the highest spirit in the world changes only too easily at the first check into depression, and one might say into a spread of fear, the French *sauve qui peut* (save himself who can). Such an army can only achieve something through its leader, never by itself. It must be led with double caution, until by degrees, in victories and hardships, its strength becomes adequate to its task. Beware then of confusing the spirit of an army with its morale.

IV

Defense

XV

OFFENSE AND DEFENSE

WHAT IS THE CONCEPTION of defense? The warding off of a blow. What, then, is its characteristic sign? The awaiting of this blow. This is the sign that makes any act defensive, and only by this sign can defense be distinguished from attack in war. But, since an absolute defense completely contradicts the conception of war, because there would then be war conducted by one side only, it follows that defense in war can only be relative, and the characteristic sign above must therefore only be applied to the conception as a whole; it must not be extended to all parts of it. A partial engagement is defensive if we await the charge of the enemy; a battle is so if we await the attack, that is, the appearance of the enemy before our position and within range of our fire; a campaign is defensive if we await the entry of the enemy into our theater of war. In all these cases the sign of awaiting and warding off the enemy attack belongs to the general conception, without any contradiction with the conception of war, for it may be advantageous to await the charge against our bayonets, or the attack on our position and our theater of war. But, because we must return the enemy's blows if we are really to carry on war on our side, the offensive action in defensive war falls in a sense under the heading of defense—that is, the offensive of which we make use falls under the conception of position, or theater of war. We can, accordingly, fight offensively in a defensive campaign, we may use some divisions in the offensive during a defensive battle, and lastly, while simply remaining in position and awaiting the enemy's attack, we still send offensive bullets into his ranks. The defensive in war is therefore not a mere shield, but a shield composed of skillfully delivered blows.

What is the object of defense? To preserve. To preserve is easier than to gain; it follows, therefore, that if the means on both sides are supposed equal, defense is easier than attack. Where, then, does the greater ease of preservation and protection lie? In this, that all the time which elapses unused falls into the scale in favor of the defender. He reaps where he has not sown. Every intermission of the attack, either from erroneous views, from fear or from indolence, favors the defender. This advantage saved Prussia from ruin more than once in the Seven Years' War. This advantage, which derives from the conception and object of the defensive, lies in the nature of all defense; it applies in other spheres of life as well as in war. In legal business, which bears much resemblance to war, it is expressed by the Latin proverb, *beati sunt possidentes.* Another advantage, arising only from the nature of war, is the assistance afforded by the lay of the land, and of this the defense has preferential use.

. . .It has been observed before in a general way that defense is easier than attack. But, since the defensive has a negative object, that of *preserving,* and the offensive a positive object, that of *conquering,* and since conquering increases our war resources but preserving does not, we must, in order to express ourselves distinctly, say that *the defensive form of war is in the abstract stronger than the offensive.* This is the result at which we have been aiming, for although it is perfectly natural and has been confirmed by experience a thousand times, it is still entirely contrary to prevalent opinion—a proof of how ideas may be confused by superficial writers.

If the defensive is the stronger form of the conduct of war, but has a negative object, it is evident that we must make use of it only so long as our weakness compels us to do so, and that we must give up that form as soon as we feel strong enough to aim at the positive object. Now, because our relative strength is usually improved if we gain a victory through the assistance of the defensive, it is also, therefore, the natural course in war to begin with the defensive and to end with the offensive. It is therefore just as much in contradiction with the conception of war to suppose the defensive the ultimate object of the war as it was a contradiction to understand passivity to belong not only to the defensive as a whole, but also to all parts of the defensive. In other words, a war in which victories are merely used to ward off blows, and where there is no attempt to return the blows, would be just as absurd as a battle in which the most passive defense should prevail in all measures.

Against the correctness of this general view, many examples might be cited of wars in which the defensive continued to be defensive to the last, and an offensive was never contemplated. Such an objection could be reasonably made only if we lost sight of the fact that here it is a question of a general conception, and that the examples which one might oppose to it are all to be regarded as cases in which the possibility of offensive reaction had not yet arrived.

. . .Having thus defined the conception of the defensive in its true meaning and established the limits of defense, we return once more to the assertion that the defensive *is the stronger form of making war.*

This will appear perfectly plain upon closer examination and comparison of the offensive and defensive, but for the present we shall confine ourselves to observing the contradiction in which the opposite view would stand with itself and with the results of experience. If the offensive form were the stronger, there would be no occasion ever to employ the defensive. Because it has, in any case, merely a negative object, everyone would necessarily want to attack, and the defensive would be an absurdity. On the other hand, it is very natural that the higher object should be purchased by greater sacrifices. Whoever considers himself strong enough to employ the weaker form may aim at the higher object; whoever aims at the lesser object can do so only in order to have the benefit of the stronger form. If we look to experience, it would probably be something unheard of if, in two theaters of war, the offensive were taken with the weaker army, and the stronger army were reserved for the defensive. But, if everywhere and at all times the reverse of this has occurred, it indicates plainly that generals still hold the defensive to be the stronger form, even though their own inclination prompts them to the offensive. . . .

V

Plan of a War

XVI

INTRODUCTION

IN THE CHAPTER ON the essence and object of war, we have sketched its general conception and pointed out its relations to surrounding things, in order to start out with a sound fundamental idea. We hinted at the many difficulties which are encountered in the consideration of this subject, and we postponed the closer examination of them and stopped at the conclusion that the overthrow of the enemy, consequently the destruction of his military forces, is the chief object of the whole act of war. This allowed us to show in the following chapter that the means which war employs is the engagement alone. In this manner we think we have obtained, for the time being, a correct point of view.

We then went separately through all the principal relations and forms which appear in military action, but are extraneous to the engagement, in order that we might determine their value more precisely, partly through the nature of the thing, partly from the experience which military history provides. We did this, moreover, in order to cleanse these relations and forms of those vague, ambiguous ideas which are generally mixed up with them, and also to bring forth in them the real object of the act of war—the destruction of the enemy's military force—as the primary object. We now return to war as a whole. We propose to discuss the plan of war and campaigns, and that obliges us to revert to the ideas in Book I.

These chapters, which are to deal with the problem as a whole, contain the very essence of strategy, in its most comprehensive and important features. We enter this innermost part of its domain, where all other threads join, with some diffidence.

Indeed, this diffidence is amply justified.

We see, on the one hand, how extremely simple the operations of war appear. We hear and read how the greatest generals speak of it in the plainest and simplest manner, how on their lips the regulation and management of this ponderous machine, with its hundred thousand parts, seems just as if it were only a question of their own persons, so that the whole tremendous act of war is individualized into a kind of duel. We find the motives for their actions explained now by a few simple ideas, now by the impulse of some emotion. We see the easy, sure—we might almost say indifferent—manner in which they treat the subject.

And now, on the other hand, we see the immense number of circumstances which present themselves for consideration by the inquiring mind; the long, often indefinite distances into which the threads of the subject spin out and the number of combinations which lie before us. If we reflect that it is the duty of theory to embrace all this systematically. . . and always to trace the action back to the necessity of a sufficient cause, there comes upon us an overpowering dread of being dragged down to pedantic dogmatism, to the regions of clumsy conceptions, where we shall never meet the great general with his easy *coup d'oeil*. If this is to be the result of an effort at theory, it would have been. . . better not to have made the attempt, for it could only bring down upon theory the contempt of genius, and would soon be forgotten. On the other hand, this easy *coup d'oeil* of the general, this simple manner of thought, this personification of the whole action of war, is the very essence of every sound conduct of war. Only in this broad way is it possible to conceive that freedom of the mind which is indispensable if the mind is to dominate events, and not be overpowered by them.

With some diffidence we proceed again; we can do so only by pursuing the way we have prescribed for ourselves from the beginning. Theory serves to cast a clear light on the mass of objects, that the mind may more easily find its bearings. Theory serves to uproot the weeds which error has everywhere sown; it shows the relationship of things to each other and separates the important from the inconsequential. Where ideas resolve themselves spontaneously into such a core of truth as is called principle. . . theory shall indicate this.

What the mind brings away with it from this journeying among the fundamental ideas of things, the rays of light that are quickened in it, *that is the assistance which theory affords*. Theory can give no formulas to solve problems; it cannot confine the mind's course to the narrow line of necessity by principles set up on both sides. It permits the mind a glimpse into the mass of objects and their relations, and then transports it again into the higher regions of action, there to act according to the measure of its natural gifts. . . to grasp the true and the right, as one single clear idea which, emerging under the combined pressure of all these forces, would seem to be a product of feeling rather than of thought.

XVII

ABSOLUTE (THEORETICAL) AND REAL WAR

THE PLAN OF A war embraces the whole military operation. Through it the operation becomes a single act, which must have one final, definitive object in which all other objects have been merged. No war is begun, or at least, no war should be begun, if people acted wisely, without first finding an answer to the question: what is to be attained by and in war? The first is the final object; the other is the intermediate aim. This dominant idea prescribes the whole course of the war, determines the extent of the means and the measure of energy; its influence manifests itself down to the smallest details of action.

We said in the first chapter that the overthrow of the enemy is the natural aim of the act of war, and that if we would remain within the strictly philosophical limits of the conception, there can be no other aim. As this idea must apply to both the belligerents, it would follow that there can be no suspension of military action, and a suspension cannot occur until one of the belligerents is actually overthrown.

In the chapter on the suspension of action in warfare, we have shown how the abstract principle of hostility, applied to its agent (man), and to all circumstances out of which war is made up, is subject to delays and limitations from causes which are inherent in the nature of the apparatus of war.

This modification, however, is not nearly sufficient to carry us from the original conception of war to the concrete form in which it almost everywhere appears. Most wars appear only as mutual anger, under the influence of which each side takes up arms to protect itself, to put fear into its adversary, and, occasionally, to strike a blow. They are, therefore, not like two mutually destructive elments brought into collision, but like voltages of two elements still apart which discharge themselves in small separate shocks.

But what is the nonconducting medium which hinders the complete discharge? Why is the philosophical conception not fulfilled? That nonconducting medium is composed of the great number of interests, forces and circumstances in the existence of the state which are affected by the war. The logical conclusion cannot be traced through their infinite windings as it would be on the simple thread of one or two inferences. In these windings it is caught fast, and man, who usually acts—in great things as well as small—more on prevailing ideas and emotions than on strictly logical conclusions, is hardly aware of his confusion, onesidedness and inconsistency.

Even if the intelligence from which war originates could have gone through all these circumstances without even momentarily losing sight of its aim, not all the other intelligences in the state which are concerned would be able to do the same. An opposition would thus arise, and a force capable of overcoming the inertia of the whole mass would become necessary—a force which will generally be inadequate to the task.

This inconsistency is found on one or the other of the two sides, or on both, and causes the war to become a half-hearted affair, without inner cohesion—something quite different from what, according to the conception of it, it should be.

This is how we find war almost everywhere, and we might doubt that our notion of its absolute nature had any reality, if we had not seen real warfare make its appearance in this absolute form right in our own time. After a short introduction performed by the French Revolution, the ruthless Bonaparte quickly brought it to this point. Under him war was carried on without slackening for a moment until the enemy was laid low, and the counter-strokes followed with almost as little remission. Is it not natural and necessary that this phenomena should lead us back to the original conception of war with all its rigorous deductions?

Shall we now rest satisfied with this and judge all wars according to it, however much they may differ from it, and deduce therefrom all the requirements of theory? We must decide upon this point, for we can say nothing intelligent concerning the plan of a war until we have decided whether war is to be only of this kind, or whether it may be of yet another kind.

If we answer the first question affirmatively, then our theory will come nearer to logical necessity in all respects; it will be a clearer and more settled thing. But what are we to say, in that case, of all the wars from Alexander, and certain of the Roman campaigns, down to Bonaparte? We would have to reject them all, and yet we could not do so without perhaps being ashamed of our presumption. The worst of it is, however, that we must say to ourselves that there may be a war of that same kind again in the next ten years, in spite of our theory, and that this theory, with its rigorous logic, is still quite powerless against the force of circumstance. We must therefore construe war not from pure conception but as it is in reality, by allowing room for everything of a foreign nature which is involved in war and attaches itself to it—all the natural inertia and friction of its parts, all of the inconsistency, the vagueness and the timidity of the human mind. We shall have to admit that war, and the form which we accord it, proceeds from ideas, emotions and circumstances prevailing for the moment. Indeed, we must admit that this has been the case even when war has taken on its absolute character; that is, under Bonaparte.

If this is the case, if we must grant that war originates and takes its form not from a final adjustment of all the innumerable relations which it affects, but from some of them which happen to predominate, it necessarily follows that it rests upon a play of probabilities, good fortune and bad, in which rigorous logical deduction often gets completely lost. It also follows that war may be a thing which is sometimes war in a greater, sometimes in a lesser, degree.

All of this theory must admit, but it is the duty of theory to place foremost the absolute form of war, and to use that form as a general point of departure, so that he who wishes to learn something from theory may accustom himself never to lose sight of it, to regard it as the fundamental standard of all his hopes and fears, in order to approach it *where he can or where he must.*

That a central idea, which lies at the root of our thoughts and actions, gives them a certain tone and character, even when the immediate reasons for a decision come from totally different regions, is just as certain as that the painter can give this or that tone to his picture by the colors which he uses for his background.

Theory is indebted to the last wars for its ability to do this effectively now. Without these warning examples of the destructive force of the unrestrained element, theory would have talked itself hoarse to no avail; no one would have believed what all have now lived to see realized.

Would Prussia have ventured to invade France in 1798 with 70,000 men, if she had foreseen that the reaction in case of failure would be so strong as to overthrow the old balance of power in Europe?

Would Prussia in 1806 have made war upon France with 100,000 men, if she had considered that the first pistol shot would be a spark to fire the mine which was to blow her into the air?

CHAPTER XVIII

INTERDEPENDENCE OF THE PARTS IN WAR

In accordance with the view of war we have—that is, whether we have in view the absolute form of war, or one of the real forms deviating more or less from it—two different notions of its result will arise.

In the absolute form, where everything is the effect of its natural and necessary cause, one thing follows another in rapid succession. There is, if we may use the expression, no neutral space. On account of the manifold reciprocal effects which war contains within itself,[1] on account of the connections in which the whole series of engagements[2] follow one after another, on account of the culminating point which every victory has, beyond which the period of losses and defeats begins—on account of all these natural circumstances of war there is, I say, only one result: the final result. Until it takes place nothing is decided, nothing won, nothing lost. Here we may say indeed: The end crowns the work. In this conception, therefore, war is an indivisible whole, the parts of which (the individual results) have no value except in their relation to this whole. The conquest of Moscow, and of half Russia in 1812, was of no value to Bonaparte unless it procured for him the peace he visualized. But it was only a part of his plan of campaign. To complete that plan, one part was still

[1] Book I, Chapter I.
[2] Book I, Chapter II.

lacking, the destruction of the Russian army. If we suppose this, added to the other success, peace was as certain as is possible for things of this kind. This second part Bonaparte could no longer attain, because he had failed to do so earlier, and so the whole of the first part was not only useless, but fatal, to him.

To this view of the connection of results in war, which may be regarded as extreme, stands opposed another extreme, according to which war is composed of single independent results, in which, as with different rounds in a game, the preceding result has no influence on those following, and we can count each separate result like separate scores made in a game.

Just as the first view derives its truth from the nature of things, so we find that of the second in history. There are innumerable cases in which it has been possible to gain a small moderate advantage without any very onerous condition being attached to it. The more the element of war is modified, the more common these cases become; but as little as the first view was ever realized in any war, just as little is there any war in which the last is true in all respects and the first can be dispensed with.

We must therefore construe war not from pure conception but as it is in reality—with all the friction of its parts and the timidity of the human mind.

If we keep to the first of these views, we must perceive the necessity of every war being looked upon as a whole from the very outset, and that at the very first step forward, the commander should have in view the end to which every line must converge.

If we admit the second view, then subordinate advantages may be pursued for their own sake, and the rest left to subsequent events.

As neither of these views is entirely without result, theory cannot dispense with either. But it makes this difference in the use of them: it requires that the first be laid down as a fundamental idea at the root of everything, and that the latter shall only be used as a modification which is justified by circumstances.

...Theory demands, therefore, that at the beginning of every war its character and main outline shall be defined according to what the political conditions and relations lead us to anticipate as probable. The more nearly, according to this probability, that its character approaches the form of absolute war, the more its outline embraces the mass of the belligerent states and draws them into the vortex—so much the more closely will its events be connected, and so much the more necessary will it also be not to take the first step without thinking what may be the last.

Of the Magnitude of the Object of the War and the Efforts to Be Made

The compulsion which we must use toward the enemy will be regulated by the proportions of our own and his political demands. Insofar as these are mutually known, they will provide the measure of the efforts to be made on each side. However, they are not always quite so evident, and this may be a first reason for a difference in the means adopted by each.

The situation and the conditions of the states are not alike; this may become a second cause.

The strength of will, the character and the capacities of the governments are as little alike; this is a third cause.

These three elements cause an uncertainty in the calculation of the resistance to be expected, consequently an uncertainty as to the extent of the means to be employed and the aim we set for ourselves.

Because insufficient efforts in war may result not only in lack of success, but also in positive loss, the two sides strive to outdo each other, which produces a reciprocal action.

This might lead to the utmost exertion of effort, if it were possible to define such a point. In this case regard for the magnitude of the political demands would be lost, the means would lose all relation to the end, and in most cases this aim at an extreme effort would be wrecked by the opposing weight of circumstances inherent in itself.

In this manner, he who undertakes war is brought back again to a middle course, in which he acts to a certain extent upon the principle of employing only such forces, and having only such a war aim in mind, as are just sufficient for the attainment of his political object. To make this principle practicable, he must renounce every absolute certainty of a result, and exclude remote contingencies from the calculation.

At this point the intellect leaves the province of strict science—of logic and mathematics—and becomes, in the widest sense of the term, art. The intellect employs the capacity to select, by instinctive judgment, the most important and decisive factors from an infinite multitude of objects and circumstances. This instinctive judgment unquestionably consists in some intuitive mental process in which the remote and unimportant facts are more quickly eliminated, and the immediate and important are sooner identified, than they could be by strictly logical deduction.

In order to ascertain the means we have to call up for the war, we must consider the political object on our own side and that of the enemy, and the power and conditions of the enemy's state and of our own. The character of the enemy government and people and the capacities of both must be weighed. We must consider the same factors on our own side, and we must take into account the political connections of other states and the effect which the war will have on those states. It is easily understood that the determination of these diverse circumstances and their diverse connections is an immense problem, that it is the true flash of genius which, confronted with them, quickly discovers the right course when it would be quite impossible to master their complexity by mere methodical study.

In this sense Bonaparte was quite right when he said that it would be a problem in algebra before which even a Newton would stand aghast.

The diversity and magnitude of the circumstances and the uncertainty as to the right course thus greatly increase the difficulty of obtaining a favorable result. We must also note that, while the overwhelming importance of the matter does not increase the complexity and difficulty of the problem, it nevertheless does increase the merit of its solution. In ordinary men freedom and activity of mind are reduced, not enhanced, by the sense of danger and responsibility; where these things give wings to strengthen the judgment, there undoubtedly must be unusual greatness of mind.

We must therefore admit that the judgment on an approaching war, on the aim which it may have, and on the means which are required, can only be formed after a general survey of all the circumstances including the most characteristic features of the moment. Next, this decision, like all in military life, can never be purely objective but must be determined by the mental and moral qualities of princes, statesmen, and generals, whether they are united in the person of one man or not.

The subject becomes general and better suited to an abstract treatment if we look at the general relations imposed upon states by their time and circumstances. We must here allow ourselves a passing glance at history.

Half-civilized Tartars, the republics of ancient times, the feudal lords and commercial cities of the Middle Ages, kings of the eighteenth century, and, lastly, princes and people of the nineteenth century, all conduct war in their own way, carry it on differently, with different means and toward different ends.

The Tartars sought new abodes. They marched out as a whole nation with their wives and children; they were, therefore, numerically stronger than any other army, and their aim was to make the enemy submit or expel him altogether. By these means they would soon have overthrown everything before them, if a high degree of civilization could have been made compatible with such a condition.

The old republics, with the exception of Rome, were of small extent, and their armies were still smaller, for they excluded the great mass of the populace. The states were too numerous and too close together not to find great enterprises obstructed by the natural equilibrium into which small separate parties always settle according to a quite general law of nature. Their wars were therefore confined to devastating open country and taking single cities in order to secure a certain influence for the future.

Rome alone is an exception to this, but not until the later period in its history. For a long time, by means of small bands, it carried on the usual warfare with its neighbors for booty and alliances. It became great more through the alliances it formed, and through which neighboring peoples gradually amalgamated with it into one whole, than through actual conquests. Only after having spread in this manner all over southern Italy did Rome begin to advance as a really conquering power. Carthage fell, Spain and Gaul were conquered, Greece subdued, and Rome's dominion extended to Egypt and Asia. At this period her military forces were immense, without her efforts being equally so. These forces were maintained by Rome's riches. Rome no longer resembled the ancient republics, or its own former self: it stood alone.

Just as distinctive in their way were the wars of Alexander. With a small army, but one distinguished for its perfection, he overthrew the rotten structures of the Asiatic states. Restlessly and ruthlessly, he penetrated the Asiatic continent as far as India. No republic could have done this. Only a king who was, so to speak, his own *condottiere,* could accomplish so much so quickly.

The great and small monarchies of the Middle Ages carried on their wars with feudal levies. Everything was restricted to a short period of time; whatever could not be done in that time was regarded as impracticable. The feudal force was raised through the organization of vassalage. The bond which held it together was partly legal obligation, partly voluntary alliance. The whole formed a real confederation. The armament and tactics were based on the right of might, on single combat, and were therefore little suited to large bodies. In fact, at no period has state unity been so weak and the individual citizen so independent. All this influenced the character of warfare during that period in the most distinctive way. Wars were carried out with comparative haste; there was little time spent idly in the field, but the object was generally only to punish, not subdue, the enemy. They carried off his cattle, burned his towns, and then returned home.

The great commercial towns and small republics introduced the *condottieri,* which was an expensive—and therefore limited—military force. In point of intensive strength, it was of still less value; it showed so little energy and strength in the field that its combats were for the most part shams. In a word, hatred and enmity no longer roused personal activity in a state, but had become articles in its trade. War had lost much of its danger, altered completely its nature, and nothing we could say of the character it then assumed would be applicable to it in reality.

The feudal system condensed itself by degrees into a definite territorial sovereignty, the internal cohesion of the state become stronger, personal obligations were transformed into material ones—money gradually becoming the substitute in most cases—and the feudal levies were transformed into armies of mercenaries. The *condottieri* formed the connecting link in the change and were therefore, for a time, the instrument of the more powerful states. This had not lasted long when the soldier, hired for a limited term, was turned into a *standing mercenary,* and the military force of states now became the standing army, supported by the public treasury.

Naturally the slow advance to this stage brought about many different combinations of the three kinds of military force. Under Henry IV we find the feudal contingents, *condottieri* and standing army all employed together. The *condottieri* existed up to the period of the Thirty Years' War; indeed, there are some slight traces of them even in the eighteenth century.

The other relations of the states of Europe at these different periods were quite as peculiar as their military forces. Europe had split up into a mass of petty states, partly republics in a state of internal dissension, partly small monarchies in which the power of the government was very limited and insecure. Such a state could not be considered as a real unity; it was rather an agglomeration of loosely connected forces. Neither, therefore, could such a state be considered an intelligent being, acting in accordance with simple logical rules.

It is from this point of view that we must look at the foreign relations and wars of the Middle Ages. Let us think of the continual expeditions of the Emperors of Germany into Italy for five centuries, without any substantial conquest of that country resulting from them or even having been intended. It is easy to look upon this as a blunder repeated over and over again as a false view which had its root in the nature of the times, but it is more reasonable to regard it as the consequence of a hundred important causes which we can partially understand, but which it is impossible for us to realize as vividly as those people could who were brought into actual contact with them. As long as the great states which have risen out of this chaos required time to consolidate and organize themselves, their whole power and energy was directed solely *to that point;* their foreign wars were few, and those that occurred bear the stamp of an incompletely developed political unity.

The wars between France and England are the first that appear, and yet at that time France is not to be considered as really a monarchy, but as an agglomeration of dukedoms and countships. England, although bearing more the semblance of unity, still fought under the feudal organization, and was hampered by many domestic troubles.

Under Louis XI, France made its greatest step toward political unity; under Charles VIII it appeared in Italy bent on conquest, and under Louis XIV it had brought its political state and its standing army to the highest perfection.

Spain attained unity under Ferdinand the Catholic. Through accidental marriage connections, the great Spanish monarchy—composed of Spain, Burgundy, Germany and Italy united—suddenly arose under Charles V. What this colossus lacked in unity and internal political cohesion it made up for in gold, and its standing army came for the first time into collision with the standing army of France. After Charles' abdication, the great Spanish colossus split into two pieces, Spain and Austria. The latter, strengthened by the acquisition of Bohemia and Hungary, now appeared on the scene as a great power, towing the German Confederation behind her.

The end of the seventeenth century, the time of Louis XIV, is to be regarded as the historical point at which the standing military power, as it existed in the eighteenth century, reached its zenith. Military power was based on voluntary enlistments and money. States had organized themselves into complete unities. The governments, by commuting the personal obligations of their subjects into taxes, had concentrated their whole power in their treasuries. Through rapid progress in social improvements, and a constantly developing system of government, this power had become very great in comparison to what it had been. France appeared in the field with a standing army of 200,000 men or more, and the other powers in proportion.

The other relations of states had likewise altered. Europe was divided into a dozen kingdoms and a few republics; it was conceivable that two of these powers might fight with each other without ten times as many others being involved, as would certainly have been the case formerly. The possible combinations in political relations were still extremely various, but they could be surveyed and determined from time to time according to probability.

Internal relations had almost everywhere been simplified into a pure monarchical form. The rights and influence of privileged estates had gradually died away, and the cabinet had become a complete unity, acting for the state in all its external relations. The time had therefore come when a suitable instrument and a despotic will could give war a form in accordance with its theoretical conception.

At this epoch appeared three new Alexanders—Gustavus Adolphus, Charles XII, and Frederick the Great—whose aim was, by means of small but highly perfected armies, to raise little states to the rank of great monarchies, and to throw down everything that stood in their way. Had they had to deal only with Asiatic states they would have more closely resembled Alexander in the parts they acted. In any case, we may look upon them as the precursors of Bonaparte with regard to what may be risked in war.

But what war gained on the one side in force and consistency was lost again on the other side.

Armies were supported out of the treasury, which the sovereign regarded partly as his private purse, or at least as a resource belonging to the government and not to the people. Relations with other states, except for a few commercial matters, concerned only the interests of the treasury or of the government, not those of the people; at least ideas tended everywhere in that direction. The cabinets, therefore, tended to look upon themselves as the owners and keepers of large estates, which they were continually seeking to increase without the tenants on these estates being particularly interested in this improvement. The people, who in the Tartar invasions were everything in war, who in the old republics and in the Middle Ages (if we properly restrict the idea to those actually possessing the rights of citizens) were of great consequence, in the eighteenth century were absolutely nothing directly, and influenced war only indirectly through their general virtues and faults.

In this manner, in proportion as the government separated itself from the people and regarded itself as the state, war became exclusively a business of the government, which it carried on by means of the money in its coffers and the idle vagabonds it could pick up in its own and neighboring countries. The consequence of this was that the means which the government could command had fairly well-defined limits, which could be mutually estimated as to both extent and duration. This robbed war of its most dangerous feature, namely the effort toward the extreme and the hidden series of possibilities connected therewith.

The financial means, the contents of the treasury and the state of credit of the enemy were all approximately known as well as the size of his army. Any large increase at the outbreak of a war is infeasible. Since the limits of the enemy's powers were thus recognized, a state felt fairly secure against complete subjugation, and as the state was conscious at the same time of the limitations of its own means, it saw itself restricted to a moderate aim. Protected from an extreme, there was no necessity to venture to an extreme. Necessity no longer giving an impulse in that direction, that impulse could now be given only by courage and ambition. These, however, found a powerful counterpoise in the circumstances of the state. Even kings in command were obliged to use the instrument of war with caution. If the army were dispersed, no new one could be obtained, and apart from the army there was nothing. This imposed the necessity for great prudence in all undertakings. It was only when a decided advantage seemed to present itself that use was made of the costly instrument; to bring about such an opportunity was the general's master stroke, but until it was brought about. . . there was no reason for action—all forces, all motives seemed to rest. The original motive of the aggressor faded away in prudence and circumspection.

Thus war became, in reality, a regular game in which time and chance shuffled the cards. In its significance, it was only a somewhat intensified diplomacy, a more forceful way of negotiating, in which battles and sieges were the diplomatic notes. Even the most ambitious aimed only at some moderate advantage in order to make use of it in peace negotiations.

This restricted, shriveled form of war proceeded, as we have said, from the narrow basis on which it rested. But the fact that excellent generals and kings, like Gustavus Adolphus, Charles XII and Frederick the Great, at the head of armies just as excellent, could not emerge more prominently from the mass of things in general—that even these men were obliged to be content to remain at the level of moderate achievement—is to be attributed to the balance of power in Europe. Now that states had become greater, and their centers farther apart, what had formerly been done through direct and perfectly natural interests—proximity, contact, family connections, personal friendship—to prevent any single state from becoming suddenly great was now effected by a higher cultivation of the art of diplomacy. Political interests, attractions and repulsions had developed into a very refined system, so that no cannon shot could be fired in Europe without all the cabinets having an interest in it.

A new Alexander had therefore to wield a good pen, in addition to his good sword, and yet he never went very far with his conquests. But although Louis XIV intended to overthrow the balance of power in Europe, and at the end of the seventeenth century had already reached such a point as to trouble himself little about the general feeling of animosity, he carried on war in the traditional manner, for while his army was certainly that of the greatest and richest monarch in Europe, it was just like the others in its nature.

Plundering and devastating the enemy's country, which played such an important part with the Tartars, with ancient nations, and even in the Middle Ages, were no longer in accordance with the spirit of the age. They were justly regarded as unnecessary barbarity, which might easily induce reprisals, and which did more injury to the enemy's subject than to his government, producing no effect and only serving to retard indefinitely the progress of national civilization. War, therefore, confined itself more and more, with regard to both means and ends, to the army itself. The army, with its fortresses and some prepared positions, constituted a state within a state, within which the element of war slowly consumed itself. All Europe rejoiced at its taking this direction, and held it to be the necessary consequence of the spirit of progress. Although there was an error in this, because the progress of the human mind can never lead to what is absurd, can never make five out of twice two, as we have already said and must repeat, still on the whole this change had a beneficial effect for the people. On the other hand, it is not to be denied that it had a tendency to make war still more an affair of the government, and to separate it still more from the interests of the people.

The plan of a war on the part of the state assuming the offensive in those times was generally the conquest of one of the enemy's provinces; the plan of the defender was to prevent this. The particular plan of campaign was to take one or another of the enemy's fortresses, or to prevent one of our own from being taken. It was only when a battle became unavoidable for this purpose that it was sought for and fought. Whoever fought a battle without this unavoidable necessity, from mere innate desire to gain a victory, was considered a daring general. Generally, the campaign was over with one siege or, if it was a very active one, with two sieges and winter quarters, which were considered a necessity. During these the faulty arrangements of the one party could never be taken advantage of by the other, and the mutual contacts of the two parties almost ceased. Winter quarters formed a distinct limit to the activity which was to take place during a campaign.

If the opposing forces were too nearly equal, or if the aggressor was decidedly the weaker, then neither battle nor siege occurred, and all the operations of the campaign pivoted on the maintenance of certain positions and magazines, and the regular devastation of particular districts of country.

As long as war was universally conducted in this way, and the natural limits of its force were so close and obvious, no contradiction was seen in it. All was considered to be normal, and criticism—which in the eighteenth century began to turn its attention to the field of the art of war—was directed to details without much concern for beginning or end. . . . Only occasionally did a penetrating judgment appear and sound common sense recognize that with superior numbers something positive must be attained or the war was being badly conducted, whatever art might be displayed.

Thus matters stood when the French Revolution broke out. Austria and Prussia tried their diplomatic art of war, which soon proved insufficient. While, according to the usual way of seeing things, all hopes were placed on a very limited military force, in 1793 such a force as no one had previously any conception of appeared. War had again suddenly become an affair of the people, and that of a people numbering thirty millions, every one of whom regarded himself as a citizen of the state. Without entering here into the details accompanying this great phenomenon, we shall confine ourselves to the results which interest us at present. By this participation of the people in war, a whole nation—instead of a cabinet and an army—entered the scale. Henceforth, the means available—the efforts which might be called forth—no longer had any definite limits, the energy with which the war might be conducted had no counterpoise, and consequently the danger for the adversary had risen to the extreme.

If the whole War of the Revolution ran its course without all this making itself fully felt and becoming quite evident; if the generals of the Revolution did not advance irresistibly to the final aim and destroy the monarchies of Europe; if the German army now and again had the opportunity of successfully resisting and for a time checking the torrent of victory, the cause lay really in that technical imperfection with which the French had to contend, which showed itself first among the common soldiers, then in the generals, and lastly—at the time of the Directory—in the government itself.

After everything had been perfected by Bonaparte, this military power, based on the strength of the whole nation, marched over Europe with such confidence and certainty that wherever it encountered the old-fashioned armies the result was never even momentarily doubtful. A reaction, however, arose in due time. In Spain, the war became of itself an affair of the people. In Austria, in 1809, the government introduced extraordinary measures, by means of reserves and Landwehr, which came nearer the end in view, and far surpassed anything that Austria had previously conceived possible. In Russia, in 1812, the example of Spain and Austria was taken as a model. The enormous dimensions of the empire, on the one hand, allowed the preparations, although too long deferred, to take effect; on the other hand, it intensified the effect produced. The result was brilliant. In Germany, it was Prussia who rose to the occasion first, made the war a national cause and, without money or credit and with its population reduced by one-half, took the field with an army twice as strong as that of 1806. The rest of Germany followed Prussia's example sooner or later, and Austria—although less energetic than in 1809—still came forward with unusual strength. Thus it was that Germany and Russia, in 1813 and 1814, appeared in France with about a million men, including all who took an active part or were killed in these campaigns.

Under these circumstances, the energy thrown into the conduct of war was quite different, and although the energy was not quite on a level of that of the French and at some points timidity was still observed, the course of the campaigns may be said, on the whole, to have been in the new style rather than the old. In eight months the theater of war was removed from the Oder to the Seine. Proud Paris had to bow its head for the first time, and the redoubtable Bonaparte lay fettered on the ground.

Thus war, by being first on the one side and then on the other an affair of the whole nation, has since Bonaparte assumed quite a new nature, or rather it has approached much nearer to its real nature, to its absolute perfection. The means then called forth had no visible limit, the limit losing itself in the energy and enthusiasm of the government and its subjects. By the extent of the means and the wide range of possible results, as well as by the powerful excitement of feeling which prevailed, energy in the conduct of war was immensely increased. The object of its action was overthrow of the enemy, and not until the enemy lay powerless on the ground was it considered possible to stop and to come to any understanding with respect to the mutual objectives of the contest.

Thus the most violent element of war, freed from all conventional restrictions, broke loose with all its natural force. The cause was the participation of the people in this great affair of state, and this participation arose partly from the effects of the French Revolution, partly from the threatening attitude of the French toward all nations.

Now, whether this will always be the case, whether all future wars in Europe will be carried on with the whole power of states, and consequently will take place only on account of great interests closely affecting the people, or whether a separation of the interests of the government from those of the people will again gradually arise, would be a difficult point to settle; least of all shall we take it upon ourselves to settle it. But everyone will agree with us, that bounds, which to a certain extent exist only in lack of awareness of what is possible, when once thrown down, are not easily built up again; and that, at least whenever great interests are in question, mutual hostility will discharge itself in the same manner as it has in our time.

We here bring our historical survey to a close, for it was not our design hastily to assign to every age some principles of the conduct of war, but only to show how each period has had its own peculiar forms of war, its own restrictive conditions and its own prejudices. Each period would, therefore, also keep its own theory of war, even if everywhere, in early times as well as in later, the inclination had existed to work out a theory on philosophical principles. The events of each age must, therefore, be appraised with due regard to the peculiarities of the time, and only he who, less through an anxious study of minute details than through an insight into the main features, can place himself in each particular age is able to understand and appreciate its generals.

But this conduct of war, conditioned by the peculiar relations of states and of military power, must nevertheless always inherently contain something quite general with which theory is above all concerned.

The period just past, in which war reached its absolute strength, contains most of what is universally valid and necessary. But it is just as improbable that wars will henceforth have this grand character as that the wide barriers which have been opened to them will ever again be completely closed. Therefore, a theory which dwells only upon this absolute war would exclude, or condemn as errors, all cases in which external influences alter the nature of war. This cannot be the object of theory, which ought to be the science of war, not under ideal, but under real, circumstances. Theory, therefore, while casting a searching, discriminating and classifying glance at objects, should always have in view the manifold diversity of causes from which war may proceed. It should, therefore, so trace out its great features as to leave room for what is required by the exigencies of the age and the moment.

Accordingly, we must say that the object which everyone who undertakes war proposes to himself, and the means which he calls forth, are determined entirely according to the particular details of his position. On that very account they will also partake of the character of the *age* and of its *general* circumstances. Lastly, *they are always subject to the general conclusions which must be deduced from the nature of war.*

XIX

AIM OF WAR MORE PRECISELY DEFINED

THE AIM OF WAR, in its conception, must always be the overthrow of the enemy. This is the fundamental idea from which we set out.

Now, what is this overthrow? It does not always necessarily imply the complete conquest of the enemy's country. If the Germans had reached Paris in 1792, there—in all human probability—the war with the revolutionary party would have been brought to an end temporarily. It was not even necessary first to defeat their armies, for those armies were not yet to be regarded as fully effective. On the other hand, in 1814 the Allies would not have gained everything by taking Paris had Bonaparte still remained at the head of a considerable army. Since his army had been largely wiped out, the capture of Paris decided everything in both 1814 and 1815. If Bonaparte in 1812, either before or after taking Moscow, had been able to destroy the Russian army completely, as he did the Austrian in 1805 and the Prussian in 1806, the possession of that capital would most probably have brought about a peace, even though an enormous area of the country remained to be conquered. . . .

In other cases the complete conquest of a country does not suffice, as in 1807 in Prussia, when the blow against the Russian auxiliary army, in the doubtful battle of Eylau, was not decisive enough, and the undoubted victory of Friedland was required to tip the scale as the victory at Austerlitz had done a year before.

We see here, too, that the result cannot be determined from general causes. The individual causes, which no one knows except those who are on the spot, and many of a moral nature which are never heard of—even the smallest features and accidents, which appear in history only as anecdotes—are often decisive. All that theory can say here is this: The main point is to keep the dominant conditions of each party in view. Out of these conditions a certain center of gravity, a center of power and movement upon which everything depends, will form itself. Against this center of gravity of the enemy the concentrated blow of all the forces must be directed.

The aim of war must always be the overthrow of the enemy.

The little always depends on the great, the unimportant upon the important, and the accidental upon the essential. This must guide our view.

. . .But whatever may be the central point of the enemy's power, against which we are to direct our operations, the conquest and destruction of his army is still the surest and most essential beginning.

Hence we think that, according to the majority of experiences, the following circumstances chiefly bring about the overthrow of the enemy:

1. Dispersion of his army, if it forms, in some degree, an effective force.
2. Capture of the enemy's capital, if it is not merely the center of state powers but also the seat of political bodies and parties.
3. An effective blow against the principal ally, if he is more powerful than the enemy himself.

We have in the past always thought of the enemy in war as a unity, which was permissible for considerations of a very general nature. But having said that the subjugation of the enemy lies in the overcoming of his resistance, concentrated in the center of gravity, we must lay aside this supposition and consider the case in which we have to deal with more than one opponent.

If two or more states combine against a third, this constitutes—from the political aspect—only one war. However, this political union also has its degrees.

The question is whether each state in the coalition possesses an independent interest in the war, and an independent force with which to prosecute it, or whether there is one among them upon whose interests and forces the others lean for support. The more the latter is the case, the easier it is to look upon the different enemies as one alone, and the more readily we can simplify our principal enterprise to one great blow. As long as this is in any way possible, it is the most thorough and complete means of success.

We would, therefore, establish it as a principle that if we can defeat all of our enemies by defeating one of them, the defeat of that one must be the aim of the war, because in that one our blows strike the common center of gravity of the whole war. There are very few cases in which this kind of conception is not admissible, and where this reduction of several centers of gravity to one cannot be made. But if this cannot be done, there is no alternative but to regard the war as two or more separate wars, each of which has its own aim. As this case presupposes the independence of several enemies, and consequently a great superiority of their combined strength, the overthrow of the enemy will be out of the question.

We now turn more particularly to the question: When is the aim to overthrow the enemy possible and advisable? In the first place, our forces must be sufficient:

1. To gain a decisive victory over those of the enemy.
2. To make the expenditure of force which may be necessary to follow up the victory to the point at which it will no longer be possible for the enemy to restore the balance.

Next, we must feel assured that our political situation is such that this result will not excite new enemies against us, who may compel us to turn away from the first enemy.

In 1806, France was able to conquer Prussia completely, in spite of the fact that in doing so she brought down upon herself the whole military power of Russia, because she was in a condition to defend herself against the Russians in Prussia.

France was able to do the same in 1808 with respect to England, but not with respect to Austria. She had to weaken herself considerably in Spain in 1809, and would have been forced to abandon completely the contest in that country had she not possessed great superiority, both physically and morally, over Austria.

These three instances must therefore be carefully studied, so that we may not lose in the last the cause which we have won in the earlier ones, and thus lose more than we gain.

In estimating the strength of forces, and what they can accomplish, the idea very often suggests itself to look upon time, by analogy with dynamics, as a factor of the forces, and to assume accordingly that half the efforts, or half the number of forces, would accomplish in two years what could only be effected in one year by the whole force united. This view, which lies at the bottom of military plans, sometimes clearly and sometimes less plainly, is completely wrong.

Like everything else on earth, a military operation requires its time—one cannot walk from Vilna to Moscow in eight days—but there is no trace in war of a reciprocal action between time and force such as is found in dynamics.

Time is necessary to both belligerents, and the only question is: Which of the two, judging by his position, has most reason to expect *special advantages* from time? Now, excluding peculiarities on one side or the other, the vanquished plainly has the most reason, according to psychological, rather than dynamic, laws. Envy, jealousy, anxiety and perhaps even magnanimity are the natural intercessors for the unfortunate; on the one hand they create friends for him, and on the other they weaken and dissolve the coalition of his enemies. Therefore, delay is more likely to produce advantageous events for the conquered than for the conqueror. Furthermore, we must recall that to make proper use of a first victory, a great expenditure of force is necessary, as we have already shown. This expenditure is not merely to be made once, but it must be maintained, like a large household. The forces which have been sufficient to give us possession of a province are not always sufficient to meet this additional outlay. Gradually the strain upon our resources become greater, until finally our capacity is exceeded. Thus, time may, of itself, bring about a change.

. . .But if the conquered provinces are sufficiently important, if there are in them points which are essential to the well-being of the unconquered parts, so that the evil eats onward of itself, like a cancer, then it is possible that the conqueror—although nothing further is done—may gain more than he loses. In this case, if no help comes from without, time may complete the work thus begun; what remains unconquered may, perhaps, fall of itself. Thus time may also become a factor of the conqueror's forces, but this can occur only if a return blow from the conquered is no longer possible, a change of fortune in his favor no longer conceivable. This can only occur, however, when this factor of his forces is no longer of any value to the conqueror, for he has already accomplished the chief object, the danger of the crisis is past and, in short, the enemy has already been overthrown.

Our object in the above reasoning has been to show clearly that no conquest can be finished too soon, that spreading it over a *greater period of time* than is absolutely necessary for its completion, instead of *facilitating* it, makes it more *difficult.* If this assertion is true, it is also true that if we are strong enough to effect a certain conquest, we must also be strong enough to do it in one march without intermediate stations. Of course, we do not mean by this without the insignificant halts to concentrate the forces and make other necessary arrangements.

By this view, which makes the character of a speedy and uninterrupted effort toward a decision essential to offensive war, we think we have completely set aside the grounds for *that* theory which, in place of the unrestrained and continuous following up of victory, would substitute a slow and so-called methodical system as being far more sure and prudent. But even for those who have readily followed us thus far, our assertion has, after all, so much the appearance of a paradox that we consider it advisable to examine more closely the apparent reasons which are against us.

It is certainly easier to reach a nearby object than one at a distance, but when the near one does not suit our purpose it does not follow that a pause, a resting point, will enable us to get over the second half of the road more easily. A short jump is easier than a long one, but no one on that account, wishing to cross a wide ditch, would first jump into the middle of it.

If we examine closely the foundation of the conception of the so-called methodical offensive war, we shall find that it consists generally of the following things:

1. Conquest of those of the enemy's fortresses which we meet.
2. Accumulation of the necessary supplies.
3. Fortifying important points, such as *magazines, bridges, positions,* and so forth.
4. Resting the troops in winter quarters or rest and recuperation quarters.
5. Awaiting the reinforcements of the ensuing year.

If we make a formal halt, a resting point, in the offensive action in order to attain these objects, it is supposed that we gain a new base of operations and renewed strength, as if our own state were following up in the rear of the army, and as if the latter acquired renewed vigor with every new campaign.

All these praiseworthy motives may make offensive war more comfortable, but they do not make its results more certain, and are mostly only pretenses to cover certain counteracting forces, such as the temperament of the commander or the indecision of the cabinet. We shall try to roll them up from the left flank:

1. The awaiting of reinforcements is just as much to the enemy's advantage as to our own, and we may say even more to his advantage. Besides, it lies in the nature of the thing that a state can place in line nearly as many combatant forces in one year as in two, for the increase in combatant forces in the second year is trifling compared to the whole.
2. The enemy rests at the same time that we do.
3. The fortification of towns and positions is not the work of the army, and therefore is no ground for a delay.
4. In the present system of subsisting armies, magazines are more necessary when the troops are in quarters than when they are advancing. As long as we advance with success, we continually acquire some of the enemy's provision depots, which assist us when the country is poor.
5. The capture of the enemy's fortresses cannot be regarded as a suspension of the attack; it is an intensified progress, and therefore the seeming suspension which it causes is not properly a case such as we allude to; it is neither a suspension nor a lessening of the use of force. But whether a regular siege, blockade, or a mere observation of one or the other is most purposeful is a question which can only be decided according to particular circumstances. We can only say this in general: that the answer to this question can only be decided by the further question, whether by mere blockading and further advance we would not be taking too great a risk. Where this is not the case, and when there is ample room to extend our forces, it is better to postpone the formal siege till the termination of the whole offensive movement. We must therefore take care not to err by neglecting the essential by the idea of making secure that which is already conquered.

No doubt it seems as if, by further advance, we risk again what has already been won. Our opinion is, however, that no pause, no resting point, no intermediate stations are in accordance with the nature of offensive war, and that when they are unavoidable, they are to be regarded as an evil which makes the result not more certain but less certain. Further, keeping strictly to the general truth, if weakness or any other cause has obliged us to stop, a second attempt at the object we have in view is generally impossible. If such a second attempt is possible, then the stoppage was unnecessary. Finally, when an object at the very beginning is beyond our strength, then it will always remain so.

We say that this appears to be the general truth, by which we only wish to eliminate the idea that time of itself can do something for the advantage of the assailant. However, as political relations may change from year to year, on that account alone many cases may happen which are exceptions to this general truth.

It may perhaps appear that we have lost our general point of view, and have nothing in sight except offensive war, but this is not at all the case. Certainly, he who can make the complete overthrow of the enemy his object will not easily be reduced to take refuge in the defensive, the immediate object of which is only to preserve. But we must maintain throughout that a defensive without any positive principle is a contradiction in strategy as well as in tactics, and we therefore come back to the fact that every defensive, according to its strength, will seek to change to the offensive as soon as it has exhausted the advantages of the defensive. However great or small the defense may be, we must also include in it, if possible, the overthrow of the enemy. . . which is to be considered as the proper object of the defensive. We say that there are cases in which the assailant, notwithstanding the fact that he has in view such a great object, may still prefer at first to employ the defensive form. That this idea is founded in reality is easily shown by the campaign of 1812. The Emperor Alexander, in engaging in the war, perhaps did not think of destroying his enemy completely, as was later done. But would such an idea have been impossible? Moreover, having such an object, would it not still have been very natural for the Russians to begin the war on the defensive?

XX

AIM OF WAR MORE PRECISELY DEFINED (CONTINUED)

Limited Aim

In the preceding chapter we have said that by the expression "overthrow of the enemy" we understand the real absolute aim of the act of war. We shall now consider what remains to be done when the conditions under which this aim might be attained do not exist.

These conditions presuppose a great physical or moral superiority, or a great spirit of enterprise, a predilection for great risks. Now, where all this is not forthcoming, the aim of war can only be of two kinds: either the conquest of some small or moderate portion of the enemy's country, or the defense of our own until better times. This last is the usual case in defensive war.

Whether the one or the other of these aims is right in a given case can always be settled by calling to mind the expression used in reference to the last. The *waiting till more favorable times* implies that we have reason to expect such times hereafter, and this waiting—that is, defensive war—is always based on this prospect. On the other hand, offensive war—that is, taking advantage of the present moment—is always imperative when the future holds out a better prospect to our enemy than to ourselves.

The third case, which is probably the most common, is when neither party has anything definite to expect from the future, and it therefore furnishes no motive for decision. In this case offensive war is plainly imperative for him who is politically the aggressor—that is, who has the positive motive—for he has taken up arms with that object, and every moment which is lost without good reason is so much lost time *for him*.

We have here decided for offensive or defensive war on grounds which have nothing to do with the relative strength of the belligerents, and yet it might appear much more natural to choose the offensive or defensive chiefly on the basis of relative strength. However, our opinion is that just in so doing we should leave the right road. The logical correctness of our simple argument no one will dispute; we shall now see whether in concrete cases it leads to an absurdity.

Let us suppose a small state which is involved in a conflict with a very superior power, foresees that with each year its position will become worse. Should it not, if war is inevitable, make use of the time when its situation is at its relative best? It must therefore attack, not because the attack *in itself* assures any advantages—it will rather increase the disparity of forces—but because such a state is under the necessity of either bringing the matter to a final issue before the worst time arrives, or of gaining in the meantime at least some advantages by which it may maintain itself. This theory cannot appear absurd. But if this small state were quite certain that the enemy would advance against it, then certainly it can and may employ the defensive against its enemy to obtain an initial advantage. There is then, at any rate, no danger of losing time.

Further, if we suppose a small state to be at war with a greater, and the future to have no influence on their decisions, if the small state is politically the assailant, we still demand of it that it should advance toward its object.

If the small state has the audacity to propose to itself the positive object against its more powerful opponent, then it also must act; that is, it must attack, if its foe does not save it the trouble. Waiting would be an absurdity, unless at the moment of execution it has altered its political decision, a case which very frequently occurs, and contributes in no small way to giving war an indefinite character, a fact that perplexes the philosopher.

Our consideration of the limited aim leads to offensive war with such an aim, and to defensive war. We shall consider both in special chapters, but first we must turn our attention elsewhere.

Hitherto we have deduced the modification of the aim in war solely from intrinsic reasons. We have considered the nature of the political intention only in so far as it is or is not directed at something positive. Everything else in the political intention is fundamentally something extraneous to war. However, in Book I, Chapter 2, "End and Means in War", we have already admitted that the nature of the political object, the extent of our own or the enemy's demand, and our whole political relation have in reality a most decisive influence on the conduct of war. We shall therefore devote the following chapter to that subject exclusively.

XXI

INFLUENCE OF THE POLITICAL OBJECT ON THE MILITARY OBJECT

WE NEVER FIND THAT a state joining in the cause of another state takes it as seriously as its own cause. An auxiliary army of moderate strength is sent; if it is not successful, then that ally looks upon the affair as in a manner ended, and seeks to get out of it on the cheapest possible terms.

In European politics it is an established thing for states to pledge themselves to mutual assistance by an offensive and defensive alliance. This pledge is not binding to the extent that one shares in the interests and quarrels of the other. It goes only so far as a mutual promise beforehand to provide a fixed, generally very moderate, contingent of troops, without regard to the object of the war or the extent of efforts made by the enemy. In a treaty of alliance of this kind, the ally does not consider himself as engaged with the enemy in a war, properly speaking, which would necessarily have to begin with a declaration of war and end with a peace treaty. Still, this idea is nowhere fixed with any distinction, and usage varies.

The thing would have some consistency and the theory of war would have less difficulty in regard to it if this promised contingent of ten, twenty or thirty thousand men were handed over entirely to the state engaged in war, so that it might be used as required. . . . The usual practice is widely different. Usually the auxiliary force has its own commander, who depends only on his own government, which prescribes to him an object such as best suits the half-hearted measures it has in view.

Even if two states really go to war with a third, they do not always both look in like measure upon this common enemy as one that they must destroy or be destroyed themselves. The affair is often settled like a commercial transaction; each, according to the amount of the risk he incurs or the advantage to be expected, takes a share in the concern to the extent of 30,000 to 40,000 men, and acts as if he could lose no more than the amount of his investment.

This point of view is taken not only when a state comes to the aid of another in a cause in which it has only minor concern, but even when both have a common and great interest at stake. Nothing can be done without diplomatic support, and the contracting parties usually agree only to furnish a small stipulated contingent, in order to employ the rest of their military forces on the special ends to which policy may happen to lead them.

This way of regarding wars entered into because of alliances was quite general, and was obliged to give way to the natural point of view only in very modern times, when the most extreme danger drove men's minds into natural pathways (as against Bonaparte) and when boundless power compelled them to do it (as under Bonaparte). It was an abnormal thing, an anomaly, for war and peace are ideas which fundamentally can have no gradations. Nevertheless, it was no mere diplomatic tradition which reason could disregard, but deeply rooted in the natural weaknesses and limitations of human nature.

Finally, even in wars conducted without allies, the political cause of the war has a great influence on the manner in which it is conducted.

If we wish from the enemy only a small sacrifice, we are satisfied with winning only a small equivalent by means of war, and we hope to attain that with moderate efforts. The enemy reasons in very much the same way. Now, if one or the other finds that he has deceived himself in his calculation—that, instead of being slightly superior to his enemy, as he supposed, he finds himself, if anything, slightly weaker—still at that moment money and all other means, as well as sufficient moral impulse for great exertions, are very often deficient. In such a case he manages as best he can and hopes for favorable events from the future, although he has not the slightest foundation for any such hope. The war, in the meantime, drags itself feebly along, like a body worn out with sickness.

Thus it comes to pass that the reciprocal action, the effort to outdo, the violence and force of war lose themselves in the stagnation of weak motives, and that both parties move with a certain measure of security in very circumscribed spheres.

If this influence of the political object on war is once permitted, as it must be, there is no longer any limit to it, and we must tolerate descending to such warfare as consists in a *mere threatening of the enemy* and in *negotiating*.

It is evident that the theory of war, if it is to be and remain a philosophical study, finds itself in difficulty here. All that is inherent in the concept of what is essential to war seems to flee from it, and theory is in danger of being left without any point of support. The natural explanation, however, soon appears. According as a modifying principle gains influence over the act of war, or rather, the weaker the motives for action become, so much the more passive resistance is substituted for action, so much the less occurs, and the less it requires guiding principles. All military art then becomes mere prudence, the principal object of which will be to prevent the teetering balance from suddenly turning to our disadvantage, and the half-hearted war from turning into a real one.

War As an Instrument of Policy

We have considered, from various aspects, the state of antagonism in which the nature of war stands in relation to the other activities of men individually and in a social group—an antagonism which is founded in our own nature, and which therefore no philosophy can unravel—in order not to neglect any of the opposing elements. We shall now look for that unity which, in practical life, these antagonistic elements attain by partly neutralizing each other. We would have brought forward this unity at the very beginning if it had not been necessary to emphasize these very contradictions, and also to look at the different elements separately. Now, this unity is *the conception that war is only a part of political intercourse, therefore by no means an independent thing in itself.*

We know, of course, that war originates through the political intercourse of governments and nations, but it is generally supposed that such intercourse is terminated by war, and that a totally different state of things ensues, subject to no laws but its own.

We maintain, on the contrary, that war is nothing but a continuation of political intercourse with an admixture of other means. We say "with an admixture of other means" in order hereby to maintain that this political intercourse does not cease through the war itself. It is not changed into something quite different, but in its essence it continues to exist, whatever may be the means it uses. The main lines along which the events of the war proceed and to which they are bound are only the general features of policy which run all through the war until peace takes place. And how can we conceive it to be otherwise? Does the cessation of diplomatic notes stop the political relations between different nations and governments? Is not war merely another kind of writing and language for their thought? It has, to be sure, its own grammar, but not its own logic.

Accordingly, war can never be separated from political intercourse. If, in the consideration of the matter, this is done in any way, all the threads of the different relations are. . . broken, and we have before us a senseless thing without an object.

This viewpoint would be indispensable if war were entirely war, entirely the unbridled element of hostility. All the circumstances on which it rests, and which determine its leading features—our own power, the enemy's power, allies on both sides, the characteristics of the governments and their people, and so forth, as enumerated in

Book I, Chapter I—are they not of a political nature, and are they not so intimately connected with the whole political intercourse that it is impossible to separate them from it? This view is doubly indispensable if we reflect that real war is no such consistent effort tending to an extreme, as it should be according to its abstract conception, but a half-hearted thing, a self-contradiction. As such, it cannot follow its own laws, but must be looked upon as part of another whole—and this whole is policy.

Policy in making use of war avoids all those rigorous conclusions which proceed from its nature. Policy troubles itself little about final possibilities, confining its attention to immediate probabilities. If there ensues much uncertainty in the whole transaction because of this, if war thereby becomes a sort of game, the policy of each cabinet relies upon the confident belief that in this game it will surpass its opponent in skill and vision.

Thus policy makes out of the all-overpowering element of war a mere instrument. Policy changes the fearsome battle-sword, which should be lifted with both hands and the whole power of the body to strike once only, into a light, handy weapon, which is sometimes nothing more than a rapier which it uses in turn for thrusts, feints, and parries.

Thus the contradictions in which man, naturally timid, becomes involved by war may be solved, if we choose to accept this as a solution.

If war belongs to policy, it will naturally assume the character of policy. If policy is grand and powerful, so also will be war, and this may be carried to the extreme at which war attains *its absolute form*.

In this way of viewing it, therefore, we need not lose sight of the absolute form of war; rather its image must constantly hover in the background.

Only through this way of conceiving it does war become once more a unity; only thus can we regard all wars as things of one kind; only thus can judgment obtain the true and exact basis and viewpoint from which great plans may be formed and judged.

It is true that the political element does not penetrate deeply into the details of war. Sentries are not posted, patrols are not sent out on political considerations. However, it is for this reason that the political element is all the more decisive in regard to the plan of a whole war, or campaign, or often even of a battle.

For this reason we were in no hurry to establish this view at the beginning. It would have given us little help while we were engaged in particulars, and would even have distracted our attention to a certain extent. In the plan of a war or campaign it is indispensable.

There is, on the whole, nothing more important in life than to find out exactly the viewpoint from which things must be regarded and judged, and then to keep to that viewpoint, for we can only grasp the mass of events in their unity from one standpoint. It is only adherence to one point of view that can save us from inconsistency.

If, therefore, in drawing up a plan of war, it is not permissible to have two or three points of view, from which things may be regarded, now with a soldier's eye, now with an administrator's, now with a politician's, and so on, then the next question is whether *policy* is necessarily paramount and all else is subordinate to it.

It is assumed that policy unites and reconciles all the interests of internal administration as well as those of humanity and whatever else are rational subjects of consideration, for policy is nothing in itself but a mere representative of all these interests toward other states. That policy may take a wrong direction, and prefer to promote ambitious ends, private interests or the vanity of rulers does not concern us here, for under no circumstances can the art of war be regarded as policy's tutor. We can only regard policy here as the representative of all the interests of the whole community.

The sole question, therefore, is whether in forming plans for a war the political point of view should give way to the purely military (if such a viewpoint were conceivable), that is to say, should disappear altogether or subordinate itself to the military view, or whether the political must remain the ruling point of view and the military be subordinated to it.

That the political point of view should end completely when war begins would only be conceivable if wars were struggles of life or death, from pure hatred. In reality, wars are, as we said before, only the expressions or manifestations of policy itself. The subordination of the political point of view to the military would be unreasonable, for policy has created the war; policy is the intelligent faculty, war only the instrument, and not the reverse. The subordination of the military point of view to the political is, therefore, the only thing which is possible.

It has been said in the third chapter of this book, *that every war should be understood according to the probability of its character and its leading features as they are to be deduced from the political forces and conditions. Often*—indeed we may safely affirm, in our days, *almost always*—war is to be regarded as an organic whole, from which the single members cannot be separated. Every individual activity therefore flows into the whole and also has its origin in the idea of this whole. [Considering these points], it becomes perfectly clear and certain that the highest standpoint for the conduct of war, from which its leading features proceed, can be no other than that of policy.

From this point of view our plans come out as from a mold; our comprehension and judgment become easier and more natural, our convictions gain force, motives are more satisfactory and history is more intelligible.

At all events, from this point of view there is no longer a natural conflict between the military and political interests, and where it does appear, it is to be regarded merely as imperfect knowledge. That policy makes demands of war which it cannot fulfill would be contrary to the presupposition that it knows the instrument it is going to employ; contrary, therefore, to a presupposition that is natural and indispensable. But if policy judges correctly of the course of military events, it is entirely its own affair to determine the events, and the direction of events, most favorable to the aim of the war.

In a word, the art of war in its highest point of view is policy, but of course a policy which fights battles instead of writing notes.

According to this view, it is impermissible and even harmful to leave a great military event, or the plan for such an event, to *purely military judgment.* Indeed, it is unreasonable to consult professional soldiers on the plan of war that they may give a *purely military opinion,* as cabinets frequently do. Still more absurd is the demand of theorists that a statement of the available means of war should be laid before a general, so that he may draw up a purely military plan for the war or for the campaign in accordance with them. General experience teaches us that, in spite of the great diversity and development of the present system of war, the main outlines of a war have always been determined by the cabinet; that is, by a purely political and not a military organ.

This is perfectly natural. None of the principal plans which are necessary for a war can be made without an insight into the political conditions, and when people speak, as they often do, of the harmful influence of policy on the conduct of the war, they really say something very different from what they intend. It is not this influence, but the policy itself, which should be faulted. If the policy is correct, that is, if it achieves its end, it can only influence the war favorably—in the sense of that policy. If the influence of policy causes a divergence from the object sought, the cause is to be sought in a mistaken policy.

It is only when policy promises itself a wrong effect from certain military means and measures, an effect inconsistent with their nature, that it can exercise a harmful effect on war by the course it charts. Just as a person, speaking a language which he has not fully mastered, sometimes says what he does not intend, so policy will often order things which do not correspond to its own intentions.

This has happened innumerable times, and it shows that a certain knowledge of military affairs is essential to the management of political intercourse.

But before going further, we must guard ourselves against a mistaken interpretation which readily suggests itself. We are far from holding the opinion that a war minister, buried in official papers, a learned engineer or even a soldier well-seasoned in the field would, any of them, necessarily make the best minister of state in a country where the sovereign does not act for himself. In other words, we do not mean to say that this acquaintance with the nature of war is the principal qualification for a minister of state. A remarkable, superior mind and strength of character are the principal qualities he must possess; a knowledge of war may be supplied in one way or another. . . .

If war is to be entirely consistent with the intentions of policy, and policy is to accommodate itself to the means available for war—in a case where the statesman and the soldier are not combined in one person—there is only one satisfactory course left, which is to make the commander-in-chief a member of the cabinet, that he may take part in its important councils and decisions. This is, however, possible only when the government itself is near the theater of war, and things can be settled without serious waste of time.

This is what the Emperor of Austria did in 1809, and the allied sovereigns in 1813, 1814, and 1815, and the arrangement proved perfectly satisfactory.

The influence in the cabinet of any military man except the commander-in-chief is extremely dangerous; it very seldom leads to sound vigorous action. . . .

We shall now conclude with some reflections derived from history.

In the last decade of the past century, when that remarkable change took place in the art of war in Europe, by which the best armies saw a part of their method of war become ineffective, and military successes far beyond any previous conception were brought about, it certainly seemed that the art of war was to be charged with an erroneous calculation of everything. It was plain that, while confined by habit within a narrow circle of concepts, Europe had been surprised by possibilities which lay outside this circle but not outside the nature of things.

Those observers who took the most comprehensive view ascribed the circumstance to the general influence which policy had exercised for centuries over the art of war, to its very great disadvantage, and which had reduced it to a half-hearted affair, often into mere sham fighting. They were right as to the fact, but wrong in attributing it to something accidental and avoidable.

Others thought that everything was to be explained by the momentary influence of the policy of Austria, Prussia, England, etc. . . .

But is it true that the real surprise by which men's minds were seized was due to something in the conduct of war, and not rather to something in policy itself? That is: did the misfortune proceed from the influence of policy on the war or from an intrinsically wrong policy?

The tremendous effects of the French Revolution abroad were evidently brought about much less by new methods and views introduced by the French in the conduct of war than by the change in the character of statecraft and civil administration, in the character of government, in the condition of the people, and so forth. The other governments took a mistaken view of all these things; their endeavor to hold their own, with their ordinary means, against these forces of a novel kind and overwhelming strength, was a blunder in policy.

Would it have been possible to perceive and correct these errors from the standpoint of a purely military conception of war? Impossible. For if there had been a philosophical strategist, who merely from the nature of the hostile elements had foreseen all the consequences, and prophesied remote possibilities, still it would have been quite impossible for such a wholly theoretical argument to produce a practical result.

Only if policy had risen to a just appreciation of the forces which had been awakened in France, and of the new relations in the political state of Europe, could it have foreseen the consequences which were bound to follow in respect to the great features of the war. Only in this way could it have developed a correct view of the extent of the means required, and the best use to make of them.

We may therefore say that the twenty years of victory of the Revolution are chiefly attributable to the mistaken policy of the governments by which it was opposed.

It is true that these errors were first exposed in the war, and that the events of the war completely disappointed the expectations which policy entertained. But this did not occur because policy failed to consult its military advisers. The art of war in which the politician of the day could believe, namely that derived from the reality of war at the time, that which belonged to the policy of the day, that familiar instrument which policy had hitherto used—*that* art of war, I say, was naturally involved in the same error as policy, and therefore could not teach it better. It is true that war itself has undergone important alterations, in both its nature and forms, which have brought it nearer its absolute form, but these changes were not the result of the French government's having freed itself from the leading-strings of policy. They arose from an altered policy which proceeded from the French Revolution not only in France but in the rest of Europe as well. This policy had called forth other means and other powers, by which it became possible to conduct war with a degree of energy which was otherwise inconceivable.

The conduct of war is therefore policy itself, which takes up the sword in place of the pen.

Further, the actual changes in the art of war are a consequence of alterations in policy, and—far from being an argument for the separation of the two—they are, on the contrary, very strong evidence of the intimacy of their connection.

Therefore, once more: war is an instrument of policy; it must necessarily bear the character of policy, it must measure with policy's measure. The conduct of war, in its major features, is therefore policy itself, which takes up the sword in place of the pen, but does not on that account cease to think according to its own laws.

XXII

LIMITED AIM—OFFENSIVE WAR

Even if the complete overthrow of the enemy cannot be the object, there may still be one which is directly positive, and this positive object can be nothing else than a conquest of a part of the enemy's country.

The use of such a conquest is as follows. We reduce the enemy's resources generally, and consequently his military power, while we increase our own. We therefore carry on the war, to a certain extent, at his expense. Further, in negotiations for peace the possession of the enemy's provinces may be regarded as a net gain, because we can either keep them or exchange them for other advantages.

This view of the conquest of the enemy's provinces is very natural, and would be open to no objection if it were not for the fact that going from the offensive to the defensive (to hold the enemy's provinces) may often cause uneasiness. In the chapter on "The Culminating Point of the Attack" (Book VII, Chapter V)* we have sufficiently explained the manner in which such an offensive weakens the military forces, and that it may be followed by a situation which causes apprehension regarding the future.

* In the referenced chapter, which has been omitted from this edition, and in Chapters IV and XXII of Book VII, Clausewitz sets forth and discusses in detail a number of factors, well understood today, which cause a weakening of the offensive as it proceeds. The factors cited by Clausewitz include: the need to detach security forces to defend key points on the lines of communi-

The weakening of our military force by the conquest of a part of the enemy's territory has its degrees, and these depend chiefly on the geographical position of this portion of territory. The more it is an annex of our own country, being contiguous or within it, and the more it lies in the direction of our principal force, by so much the less will it weaken our military force. . . . If, on the other hand, the conquered territory is a strip running between hostile provinces and has an eccentric position and unfavorable configuration of ground, the weakening increases so visibly that a victorious battle becomes not only much easier for the enemy, but may even become unnecessary. . . . Therefore, the selection of such an object depends on our expected ability to hold possession of the conquest, on the expectation that a temporary occupation would justify the required expenditure of force, and on whether we must not anticipate a counterstroke so vigorous as to completely destroy the balance of forces.

There is just one point which we must yet add.

An offensive of this kind will not always compensate us for what we lose upon other points. While we are engaged in making a partial conquest, the enemy may be doing the same at other points, and if our enterprise is not of superior importance it will not compel the enemy to abandon his. It is, therefore, a question for serious consideration whether we do not lose more than we gain in a case of this kind.

Even if we suppose two provinces (one on each side) to be of equal importance, we shall always lose more by the one which the enemy takes from us than we can gain by the one we take, because a number of our resources. . . become ineffective. But since the

cation through conquered enemy territory, casualties sustained through battle and sickness, increasing length of lines of communication, and relaxation of efforts. He points out that, as a result of these factors and possible others, there is often a culminating point in the offensive, beyond which it is more dangerous to proceed than it is to revert to the defensive and await the enemy's counteroffensive.

same thing occurs on the enemy's side, one would suppose that in reality there is no ground to attach more importance to the holding of what is our own than to the conquest. Yet, it is so. Retaining possession of our own country is always a matter that concerns us more deeply, and the suffering inflicted on our own state cannot be outweighed or neutralized by what we gain in return, except when the latter promises a much greater advantage.

The consequence of all this is that, in a strategic attack with only a modest object, it is much more necessary to defend points which the attack does not directly cover than in an attack which is directed at the center of gravity of the enemy's power. Consequently, in such an attack, the concentration of forces in space and time cannot be carried out to the same degree. In order that it may occur, at least with respect to time, it becomes necessary for the offensive advance to be made simultaneously from every suitable point. This attack therefore loses the advantage of being able to employ a much smaller force by remaining on the defensive at particular points. In this way the effect of aiming at a minor object is to bring all things more to a single level. The whole act of war can no longer be concentrated into one principal action which can be conducted according to main points of view; it is more dispersed, the friction becomes greater everywhere, and there is everywhere more room for chance.

This is the natural tendency of the thing. The commander is weighed down by it, finds himself more and more neutralized. The more he is conscious of his own powers, the greater his determination and the resources under his command, the more he will be able to free himself from this tendency in order to give a preponderant importance to some one point, even if that is possible only by incurring greater risks.

XXIII

LIMITED AIM—DEFENSE

THE ULTIMATE AIM OF defensive war can never be an absolute negation, as we have previously observed. Even for the weakest, there must be some means by which he can hurt his opponent and some point at which he can threaten him.

We could undoubtedly say that this object can be the exhaustion of the enemy, for since the enemy has a positive object, every one of his undertakings which fails, even if it has no other effect than the loss of the force employed, may still be considered as *fundamentally* a retrograde step, while the loss which the defender suffers is not in vain, because his object of retaining possession has been achieved. This would be tantamount to saying that the defender has his positive object in merely keeping possession. Such reasoning might be valid if it were certain that the assailant, after a number of fruitless attempts, must be worn out and desist from further efforts. But just this certainty is lacking. If we examine the actual exhaustion of forces, the defender is at a disadvantage. The attack weakens, but only in the sense that there may be a turning point; if we set aside that supposition, the weakening is certainly greater on the defensive side than on the offensive, because, in the first place, the defender is weaker, and therefore if losses on both sides are equal, he loses more relatively than the attacker. In the second place, he is generally deprived of a part of his territory and resources. Accordingly, we can deduce no reason for the attacker to relax his efforts, and we can only conclude that if he repeats his blows while the defender does nothing but ward them off, the latter cannot develop any counterweight to the risk that, sooner or later, one of these attacks may succeed.

Although in reality the exhaustion, or rather the wearing down, of the stronger has often produced a peace, that fact is attributable to the half-heartedness with which war is usually conducted, and cannot be regarded philosophically as the general and ultimate aim of any defensive whatsoever. There is, therefore, no alternative but that the defensive should find its aim in the conception of awaiting the enemy, which is, moreover, its real character. This idea includes an alteration of circumstances, an improvement of the situation. When this cannot be brought about by the defensive itself, it can only be expected by means of assistance coming from without. Now, this improvement from without can proceed from nothing else but a change in political relations, either in new alliances arising in favor of the defender, or in the collapse of old ones erected against him.

This is, then, the object for the defender when his weakness does not permit him to contemplate any important counterstroke. But this is not the nature of every defensive, according to the conception which we have given. According to that concept, the defensive is the stronger form of war, and because of its strength it can also be employed when a strong counterstroke is intended.

These two cases must be kept distinct from the very first, as they have an influence on the defense.

In the first case, the defender's object is to keep possession of his own country as long as possible, because in that way he gains the most time, and gaining time is the only way to attain his object. The positive object, which he can in most cases attain, and which will provide him an opportunity to carry out his object in the peace negotiations, he cannot yet include in his war plan. In this state of strategic passivity, the advantages which the defender can gain at certain points consist in merely repelling separate attacks; the preponderance gained at those points he carries over to others, for he is usually hard-pressed at all points. If he has no opportunity of doing this, then there remains for him only the small advantage that the enemy will leave him alone for a time.

If the defender is not altogether too weak, small offensive operations directed. . . to a temporary advantage to cover losses which may be sustained later, invasions, diversions, or enterprises against a single fortress may have a place in this defensive system without altering its aim or essence.

In the second case, where the defensive is based upon an ultimate positive intention, it assumes a more positive character, which becomes correspondingly more positive as circumstances warrant an increasingly powerful counterstroke. In other words, the more the defensive has been voluntarily adopted in order to make the first blow more certain, the bolder may be the snares which the defender sets for his opponent. . . .

Thus a great positive success can never be obtained except through positive measures, planned not with a view to merely awaiting the enemy, but with a view toward a *decision.* In short, even on the defensive, there is no great gain to be won except by a great stake.

XXIV

PLAN OF WAR WHEN THE DESTRUCTION OF THE ENEMY IS THE OBJECT

HAVING CHARACTERIZED IN DETAIL the different aims to which war may be directed, we shall go through the organization of the whole war for each of the three separate gradations corresponding to these aims.

In conformity with all that has been said on the subject up to now, two fundamental principles will comprehend the whole plan of war and serve as a guide for everything else.

The first is: to trace the weight of the enemy's power back to as few centers of gravity as possible, or to one if it can be done; again, to confine the attack against these centers of gravity to the minimum possible number of principal undertakings, or to one if possible; finally, to keep all secondary undertakings as subordinate as possible. In a word, the first principle is *to concentrate as much as possible.*

The second principle is *to act as swiftly as possible;* therefore, to permit no delay or detour without sufficient reason.

The determination of the center of gravity of the enemy's power depends:

1. On how that power is politically constituted. If it consists of the armies of only one state, there is generally no difficulty; if it consists of allied armies, of which one is acting simply as an ally without much interest of its own, then the difficulty

is not much greater; if of a coalition for common objects, then it depends upon the cordiality of the alliance, which we have already discussed.

2. On the situation of the theater of war upon which the different hostile armies make their appearance.

If the enemy's forces are concentrated in one army in one theater of war, they constitute a real unity, and we need not inquire further. If they are in one theater of war but in different armies belonging to different powers, there is no longer absolute unity. There is, however, a sufficient interdependence of the parts for a decisive blow upon *one part* to bring down the other along with it. If the armies are posted in theaters of war adjoining each other, and are not separated by any great natural obstacles, there is a decided influence of the one upon the other; but if the theaters of war are separated by great distances or by neutral territory or natural obstacles, the influence is very doubtful and therefore improbable. If they are at opposite sides of the enemy state, so that operations directed against them must diverge on eccentric lines, then almost every trace of a connection has disappeared.

. . .It proceeds from what we have said that the conception of separated or connected hostile powers extends through all degrees of relationship. Therefore, only in the individual case can the influence which events in one theater may have on another be discovered, according to which we may afterwards settle how far the different centers of gravity of the enemy's strength may be reduced to one.

There is only one exception to the principle of directing all our strength against the center of gravity of the enemy's power; that is, if ancillary expeditions promise extraordinary advantages. It is assumed, in this case, that we have a decisive superiority which enables us to undertake such enterprises without risking too much in respect to our main objective.

. . .Thus the first consideration in a plan of war is to determine the centers of gravity of the enemy's power, and if possible to reduce them to one. The second is to unite the forces which are to be employed against this center of gravity into one great action.

Now, at this point the following grounds for dividing our forces may present themselves:

1. The original disposition of the military forces; therefore, also the situation of the states engaged in the offensive.

If the concentration of forces would occasion detours and loss of time, and the danger of advancing by separate lines is not too great, the division may be justifiable on these grounds. To effect an unnecessary concentration of forces, with great loss of time by which the freshness and rapidity of the first blow is diminished, would be contrary to the second leading principle which we have established. In all cases where there is a hope of surprising the enemy in some measure, this deserves particular attention.

The case becomes still more important if the attack is undertaken by allied states which are situated, with respect to the enemy state, side by side rather than one behind the other. If Austria and Prussia undertook a war against France, it would result in a waste of time and strength for both armies to set out from the same point, as the natural line for an army operating from Prussia against the heart of France is from the lower Rhine, and that of the Austrians is from the upper Rhine. Concentration could in this case only be effected by a sacrifice; consequently, in any particular instance, the question to be decided would be whether the need for concentration is so great that this sacrifice must be made.

2. The attack by separate lines may offer greater results. Since we are now speaking of advancing by separate lines against one center of gravity, we are therefore supposing an advance by *converging* lines. A separate advance on parallel or diverging lines comes under the heading of *secondary undertakings,* of which we have already spoken.

Now, every convergent attack, in strategy as well as in tactics, offers prospect of *great* results, for if it succeeds, the consequence is not simply a defeat of the enemy, but more or less the cutting off of the enemy's forces. The convergent attack is, therefore, always that which may produce greater results, but on account of the separation of the parts of the force, and the enlargement of the theater of war, it also involves the greater risk. It is the same here as with attack and defense; the weaker form offers a prospect of greater results.

The question, therefore, is whether the attacker feels strong enough to strive for this great result.

. . .After all these reflections, we think that although the convergent attack is in itself a means of obtaining greater results, still it should generally proceed only from a previous separation of military powers, and there will be few cases in which we should be correct in giving up the shortest and most direct line of operation in order to undertake the convergent attack.

3. The breadth of a theater of war can be motive for attacking on separate lines.

If an army on the offensive from any point advances farther into the interior of the enemy country, the space which it commands is certainly not restricted exactly to the line of road by which it marches, but it will extend to some distance on each side. That distance will depend very much upon the solidity and cohesion, if we may use that term, of the opposing state. If the hostile state is only loosely united, if its people are a soft race unaccustomed to war, then, without our taking very much trouble, a considerable extent of country will remain open behind our victorious army. If, on the other hand, we have to deal with a brave and loyal population, the space behind our army will form a more or less acute triangle.

In order to prevent this evil, the attacker desires to arrange his advance on a certain width of front. If the enemy's force is concentrated at a particular point, this breadth of front can be retained only so long as we are not in contact with the enemy, and it must be reduced as we approach his position. This is easily understood.

But if the enemy himself has taken up a position with a certain extent of front, then there would be nothing unreasonable in a corresponding extension on our part. We speak here of a single theater of war, or of several which lie near one another. Clearly this is the case when, according to our view, the chief operation is also to be decisive on subordinate points.

But, can we *always* take this chance? And may we expose ourselves to the danger which must arise if the influence of the major operation is not sufficient to decide at minor points? Does not the need of a certain breadth for a theater of war deserve special consideration?

Here as everywhere else it is impossible to exhaust the number of combinations which *may* take place, but we maintain that, with few exceptions, the decision on the main point will carry with it the decision on all minor points. Therefore, the action should be arranged in accordance with this principle in all cases in which the contrary is not evident.

. . .We. . . declare ourselves completely opposed in principle to the dependence of the chief attack on secondary points, and we maintain that an attack directed toward the destruction of the enemy which has not the boldness to shoot. . . direct at the heart of the enemy's power can never hit the mark.

4. Lastly, there is still a fourth ground for a separate advance in the ease it may afford for subsistence.

It is certainly much more pleasant to march with a small army through a rich province than with a large army through a poor one, but by suitable measures and an army accustomed to privations, the latter is not impossible. The first should therefore never have such an influence on our plans as to expose us to great danger.

We have now done justice to the grounds for a separation of forces which divides a main operation into several. If the separation occurs on any of these grounds, with a distinct conception of the object, and after due consideration of the advantages and disadvantages, we shall not venture to find fault.

But if, as usually happens, the plan is produced in this way by a learned general staff, according to mere routine; if different theaters of war, like the squares on a chess board, must each be occupied by its piece before the moves begin; if these moves approach the aim in complicated lines and relations by dint of imagined skill in combinations; if the armies are to separate today in order to apply all their skill in reuniting at the greatest risk two weeks hence—then we detest this abandonment of the direct, simple common sense road in order to plunge intentionally into sheer confusion. This folly occurs the more easily the less the commander-in-chief directs the war, and conducts it in the sense, which we have pointed out in the first chapter, as an act of his individuality vested with extraordinary powers. It occurs the more easily, therefore, the more the whole plan is concocted by an unpractical general staff and from the ideas of a dozen smatterers.

We must still consider the third part of our first principle; that is, to keep the subordinate parts as subordinate as possible.

By endeavoring to reduce all operations of the war to a *single* object, and trying to attain this as far as possible by one great action, we deprive the other points of contact of the belligerent states of a part of their independence; they become subordinate actions. If we could concentrate everything absolutely into one action, those points of contact would be completely neutralized; but this is seldom possible, and it is therefore a matter of keeping them so far within bounds that they shall not draw off too much force from the main action.

Next, we maintain that the plan of war must have this tendency even when it is not possible to reduce the whole resistance of the enemy to one center of gravity. Consequently, in case we are placed in the position, already mentioned, of carrying on two almost separate wars at the same time, the one must always be looked upon as the *main issue* to which our forces and activity are to be chiefly devoted.

According to this view, it is advisable to advance *offensively* only toward that one principal point, and to remain on the defensive at all others. Only where very exceptional circumstances invite an attack elsewhere would it be justified.

Further, we will seek to maintain this defensive, which takes place at minor points, with as few troops as possible, and to avail ourselves of every advantage which the defensive form can give.

This view applies with even greater force to all theaters of war in which, though armies belonging to different powers also appear, they are still such as will be struck when the general center of gravity is hit.

But against *that* enemy at whom the main blow is directed there can no longer, according to this view, be any defensive on minor theaters of war. The main attack itself and the secondary attacks, which are brought about by other considerations, comprise this blow, and make every defensive, on points not covered by it, superfluous. All depends on the main decision; by it every loss will be compensated. If the forces are sufficient to make it reasonable to seek such a main decision, then the *possibility of failure* can no longer be used as a ground for guarding oneself against any possibility of injury at other points; for *that very course* makes failure much more probable, and therefore introduces a contradiction into our action.

This predominance of the principal action over the minor ones must also be the principle observed even in separate branches of the whole attack. However, since there are usually other motives which determine which forces shall advance from one theater of war and which from another against the common center of gravity, we mean here only that *there must be an effort to make the principal action predominate,* for everything will become simpler and less subject to the influence of chance the more this state of predominance can be attained.

The second principle concerns the rapid use of forces.

Every unnecessary expenditure of time, every unnecessary detour, is a waste of power and therefore contrary to the principles of strategy.

It is extremely important always to bear in mind that almost the only advantage which the offensive possesses is the effect of surprise at the opening of the scene. Suddenness and irresistible impetus are its strongest elements, and when the object of the attack is the complete overthrow of the enemy, it can rarely dispense with them.

Theory demands, therefore, the shortest way to the object, and completely excludes from consideration endless discussions about right and left, here and there.

If we recall what was said in the chapter on the object of the strategic attack (Book VII, Chapter 3) regarding the weakest point of the state, and, further, what appears in Chapter 4 of this book on the influence of time, we believe that no further argument is required to prove that the influence which we claim for that principle really belongs to it.

Bonaparte never acted otherwise. The shortest main road from army to army, from one capital to another, was his favorite.

And in what will now consist the principal action to which we have referred everything, and for which we have demanded a swift and straightforward execution?

In Chapter 4 we have explained, as far as it is possible in a general way, what the overthrow of the enemy means, and it is unnecessary to repeat that explanation. Whatever it may finally depend upon, the beginning is in all cases the same: *the destruction of the enemy's military force,* that is, *a great victory over it, and its dispersion.* The sooner—which means the nearer our own frontier—this victory is sought, the *easier* it is. The later it is gained—that is, the deeper in the heart of the enemy's country—the more *decisive* it is. Here, as everywhere, the facility of success and its magnitude balance each other.

If we are not so superior to the enemy that our victory is beyond doubt, then we must, when possible, seek him out; that is, seek out his principal force. We say *when possible,* for if this endeavor to find him led to great detours, false directions, and loss of time, it might very likely turn out to be a mistake. If the enemy's main force is not on our road and our interests otherwise prevent our searching for it, we may be sure we shall meet it later, for it will not fail to place itself in our way. We shall then, as we have said, fight under less advantageous circumstances—an evil to which we must submit. If we gain the victory in spite of that, it will be so much the more decisive.

Once the great victory is gained, there should be no talk of rest, but only of pursuit.

It follows, therefore, that in the case here assumed, it would be an error to bypass the enemy's main force purposely, if it places itself in our way, at least if we expected thereby to facilitate the victory.

On the other hand, it follows that if we have a decided superiority over the enemy's main force, we may purposely bypass it in order to obtain a more decisive battle at a future time.

. . .Once the great victory is gained, there should be no talk of rest, of pausing for breath, of considering, or of consolidating and so forth, but only of pursuit, of fresh blows wherever necessary, of the capture of the enemy's capital, of attacking the enemy's auxiliary forces, of whatever else appears to be the remaining center of power of the enemy state.

. . .We demand, therefore, that the main force should press forward rapidly in pursuit, without any rest. We have already condemned the idea of allowing the advance towards the principal point being made dependent upon success at secondary points; the consequence of this is ordinarily that our main army only keeps behind it a narrow strip of territory which it can call its own, and which therefore constitutes its theater of war. How this weakens the momentum at the head, and the dangers for the offensive arising therefrom, we have shown earlier. Will not this difficulty, will not this inherent counterpoise come to a point which impedes further advance? That certainly may occur. But just as we have insisted above that it would be a mistake to try to avoid this contracted theater of war from the start, and for the sake of that object to rob the advance of its impetus, so we also now maintain, that as long as the commander has not yet overthrown his opponent, as long as he considers himself strong enough to attain that aim, so long must he also pursue it. He does so perhaps at an increased risk, but also with the prospect of a greater success. If he reaches a point beyond which he cannot venture to proceed, where he considers it necessary to protect his rear and extend his forces right and left—well, then, this is probably his culminating point. His momentum is then spent, and if the enemy is not overthrown, then it is most probable that nothing will come of it.

All that the assailant now does to intensify his attack by a slow conquest of fortresses, defiles, and provinces is no doubt still a slow advance, but it is only a relative one; it is no longer absolute. The enemy is no longer in flight, he is perhaps preparing a renewed resistance; it is therefore quite possible that, although the attacker continues to advance with all his force, the position of the defense is improving every day. In short, we come back to this, that as a rule there is no second assault after a halt has once become necessary.

We are not so foolish as to maintain that no instance can be found of states having *gradually* been reduced to the utmost extremity. In the first place, the principle we now maintain is no absolute truth, to which an exception is impossible, but one founded only on the ordinary and probable result. Next, we must make a distinction between cases in which the downfall of a state has actually been effected by a slow, gradual process, and those in which the event was the result of the first campaign. We are here only treating the latter case, for it is only in this case that there is that tension of forces which either overcomes the center of gravity of the burden, or is in danger of being overcome by it.

If we gain a moderate advantage in the first year, and add another in the following year, and thus gradually advance toward our object, there is nowhere very imminent danger, but for that reason it is distributed over many points. Each pause between successes gives the enemy fresh chances. The effects of the first successes have very little influence on those which follow, often none, often a negative one only, because the enemy recovers, or is excited to fresh resistance, or obtains foreign aid; whereas, when everything happens in one campaign, yesterday's success carries today's with it, one fire lights itself from another. If there are cases in which states have been overcome by successive blows—in which, consequently, *time,* generally the patron of the defense, has proved adverse—how infinitely more numerous are the instances in which the designs of the aggressor have by that means utterly failed! We need only think of the result of the Seven Year's War, in which the Austrians sought to attain their object with such comfort, caution, and prudence that they completely missed it.

Taking this view, we therefore cannot possibly join in the opinion that the concern for the proper arrangement of a theater of war and the impulse which urges us onward are of equal importance, and that the former must, to a certain extent, be a counterbalance to the latter. We look upon the disadvantages which spring from the forward movement as an unavoidable evil which deserves attention only when, ahead of us, there is no longer any hope for us.

. . .So much for the main operation, its necessary tendency, and its unavoidable risks. With respect to the subordinate operations, we say that there must, above all things, be a common aim for all of them, but this aim must be such as not to paralyze the action of the individual parts. If we invade France from the Upper and Middle Rhine and Holland with the intention of uniting at Paris, but with none of the armies supposed to risk anything on the advance and to keep itself intact until the concentration is effected, that is a *ruinous* plan. There must be a balancing of this threefold movement, causing delay, indecision, and timidity in the forward movement of each of the armies. It is better to assign to each part its mission, and only unity at that point where these several activities become a unity of themselves.

Therefore, when a military force advances to the attack on separate theaters of war, a separate object should be assigned to each army, against which it can direct the force of its attack. The important thing is that *this attack* should take place from all sides simultaneously, but not that all should gain proportional advantages.

. . .With regard to the personal characteristics of generals, everything in this becomes a matter of the individual. We must not, however, fail to make one general remark, which is, that we should not, as is generally done, place at the head of subordinate armies the most prudent and cautious commanders, but *the most enterprising*. In strategic operations separately conducted, there is nothing more important than that every part should develop its full powers, so that errors committed at one point may be compensated for by successes at others. This complete activity at all points is only to be expected when the commanders are spirited, enterprising men who are driven forward by an inner urge from their own hearts; an objectively, coolly reasoned conviction of the necessity for action seldom suffices.

. . .What we have said up till now concerning a plan of war in general, and in this chapter concerning particularly that which is directed to the overthrow of the enemy, was intended to give special prominence to the object of the plan, and in addition to indicate principles which shall serve as guides in the preparation of ways and means. Our desire has been to give in this way a clear conception of what one wants and ought to do in such a war. We have tried to emphasize the necessary and general, and to leave a margin for the play of the particular and accidental, but to exclude all that is *arbitrary, unfounded, trifling, fantastic or sophistic.* If we have succeeded in this object, we look upon our task as accomplished.

VI

I Believe and Profess

I BELIEVE AND PROFESS

I BELIEVE AND PROFESS that a people never must value anything higher than the dignity and freedom of its existence; that it must defend these with the last drop of its blood; that it has no duty more sacred and can obey no law that is higher; that the shame of a cowardly submission can never be wiped out; that the poison of submission in the bloodstream of a people will be transmitted to its children, and paralyze and undermine the strength of later generations; that honor can be lost only once; that, under most circumstances, a people is unconquerable if it fights a spirited struggle for its liberty; that a bloody and honorable fight assures the rebirth of the people even if freedom were lost; and that such a struggle is the seed of life from which a new tree inevitably will blossom.

I believe and profess that a people never must value anything higher than the dignity and freedom of its existence.

I declare and assert to the world and to future generations that I consider the false wisdom which aims at avoiding danger to be the most pernicious result of fear and anxiety. Danger must be countered with virile courage joined with calm and firm resolve and clear conscience. Should we be denied the opportunity of defending ourselves in this manner, I hold reckless despair to be a wise course of action. In the dizzy fear which is beclouding our days, I remain mindful of the ominous events of old and recent times, and of the honorable examples set by famed peoples. The words of a mendacious newspaper do not make me forget the lessons of centuries and of world history.

I assert that I am free of all personal ambitions; that I profess thoughts and sentiments openly before all citizens; and that I would be happy to find a glorious end in the splendid battle for the freedom and excellence of my country.

Does my faith and the faith of those who think like me deserve the contempt and scorn of our citizens? Future generations will decide.

A nation cannot buy freedom from the slavery of alien rule by artifices and stratagems. It must throw itself recklessly into battle, it must pit a thousand lives against a thousand-fold gain of life. Only in this manner can the nation arise from the sick bed to which it was fastened by foreign chains.

Boldness, that noble virtue through which the human soul rises above the most menacing dangers, must be deemed to be a decisive agent in conflict. Indeed, in which sphere of human activity should boldness come into its own unless it be in struggle?

Boldness is the outstanding military quality, the genuine steel which gives to arms their luster and sharpness. It must imbue the force from camp follower and private to the commander-in-chief.

In our times, struggle, and, specifically, an audacious conduct of war are practically the only means to develop a people's spirit of daring. Only courageous leadership can counter the softness of spirit and the love of comfort which pull down commercial peoples enjoying rising living standards. Only if national character and habituation to conflict interact constantly upon each other can a nation hope to hold a firm position in the political world.

A nation which does not dare to talk boldly will risk even less to act with courage.

A nation does not go under because for one or two years it engages in efforts which it could not sustain for ten or twenty years. If the importance of the purpose demands it, and especially if it is a matter of maintaining independence and honor, such efforts are a call of duty. The government possesses all the means required to persuade the people to live up to their obligations. It is entitled to expect exertions, to insist on them, and if necessary is bound to compel compliance. Strong and purposive governments, which are truly capable of managing affairs, never will fail to act in this manner.

Perhaps there never again will be times when nations will be obliged to take refuge in the last desperate means of popular uprising against foreign domination. Yet in our epoch, every war inevitably is a matter of national interest and must be conducted in that spirit, with the intensity of effort which the strength of the national character allows and the government demands.

In my judgment the most important political rules are: never relax vigilance; expect nothing from the magnanimity of others; never abandon a purpose until it has become impossible, beyond doubt, to attain it; hold the honor of the state as sacred.

The time is yours; what its fulfillment will be, depends upon you. . . .